ENVIRONMENTAL CHEMISTRY IN ANTARCTICA

CURRENT TOPICS IN ENVIRONMENTAL AND TOXICOLOGICAL CHEMISTRY
Edited by O. Hutzinger and J. Albaigés

ENVIRONMENTAL CHEMISTRY IN ANTARCTICA

Edited by
Paolo Cescon
Department of Environmental Sciences
University Ca'Foscari of Venice, Italy

CRC Press is an imprint of the
Taylor & Francis Group, an **informa** business

First published 2001 by OPA (Overseas Publishers Association)

Published 2021 by CRC Press
Taylor & Francis Group
6000 Broken Sound Parkway NW, Suite 300
Boca Raton, FL 33487-2742

ISBN 13: 978-90-5699-285-9 (hbk)
ISSN 0275-2581

DOI: 10.4324/9781315078267

Visit the Taylor & Francis Web site at
http://www.taylorandfrancis.com

and the CRC Press Web site at
http://www.crcpress.com

The articles published in this book first appeared in the *International Journal of Environmental Analytical Chemistry,* Volume 60, Number 1; Volume 61, Number 3; Volume 61, Number 4; Volume 62, Number 3; Volume 63, Number 1; Volume 63, Number 2; Volume 63, Number 3; Volume 66, Number 1; Volume 68, Number 4; Volume 75, Number 4.

British Library Cataloguing in Publication Data

Environmental chemistry in Antarctica. – (Current topics in environmental and toxicological chemistry ; v. 17)
1. Environmental chemistry – Antarctica 2. Pollution – Antarctica
I. Cescon, P.

CONTENTS

PREFACE

This volume is a collection of papers produced within the framework of the Italian National Antarctic Research Programme (PNRA) on the monitoring and control of environmental contamination. The volume represents a contribution of the PNRA to the study of planetary contamination and to the understanding of the processes of global change.

The research focuses on the measurement and analysis of trace elements and organic micropollutants in the following matrices: snow/firn, seawater, soils, sediments, suspended particulate matter, pack ice, atmosphere, and biota.

The results presented extend beyond the development of specific analytical methodologies, to explicitly tackle significant environmental issues concerning global changes. Particularly relevant are the results concerning time changes of CFCs in the troposphere and lead concentration in Antarctic snow in Victoria Land, the presence of organic micropollutants in various Antarctica matrices, and the seasonal evolution of trace elements and organics in seawater.

Paolo Cescon
Coordinator of the Chemical
Contamination Project – PNRA

NON-SEA-SALT CONTRIBUTION OF SOME CHEMICAL SPECIES TO THE SNOW COMPOSITION AT TERRA NOVA BAY (ANTARCTICA)

G. PICCARDI, F. CASELLA and R. UDISTI

Department of Public Health and Environmental Analytical Chemistry, University of Florence, Via G. Capponi 9, 1-50121 Florence, Italy

The concentrations of main components of two series of snow samples collected in the area of Terra Nova Bay have been determined. From statistical analysis of the obtained values, Na, Cl and Mg result from marine contribution, whereas a constant excess of K and of Ca is added to the sea spray content of snow. A further source of sulfates, besides the ones produced in DMS oxidation, is evidenced in the $nssSO_4$ contribution. A good correlation among $nssSO_4$, MSA and NO_3 or nssCa is also observed.

KEY WORDS: Snow analysis, Antarctica, snow chemical composition, main components

INTRODUCTION

The interest of studying snow composition with depth of coastal areas of Antarctica is due to two principal reasons. First, the relatively high quantity of precipitation allows to distinguish the characteristics of single events, and second the proximity to sea facilitates the bio-geochemical investigation of aerosol particles and their deposition and conservation into the snow. The principal components analysis is also useful in the attempt to individualize different sources of snow constituents. The concentration of a single component is usually divided in two contributions: sea-salt and non sea-salt. The Na and Cl concentrations can be used to obtain the marine contribution of other components assuming that Na and Cl are exclusively of marine origin and that the sea spray is of the same composition of subsurface sea-water. Some authors used concentrations of elements of crustal[1,2] or cosmic origin[3] with the attempt to give a more complete interpretation to the non marine snow contributions. Contributions of sources different from sea salt, indicated as nss or ex, are frequently evidenced from enrichment factor (E.F.) calculations obtained as:

$$E.F. = ([X]/[Na])_{snow}/([X]/[Na])_{sea}$$

Table 1 shows some E.F. values obtained from the literature and usually calculated from mean concentration values and those we recalculated from the mean concentration of different papers using the above cited formula.

Table 1 Enrichment factors (E.F.) of Mg, K and Ca in snow samples of Antarctica.

Site	*Latitude*	*Longitude*	*alt/m*	*dist/Km*	*EF/Mg*	*EF/K*	*EF/Ca*	*ref*
Byrd station	79°59'S	120°01'W	1530	700	1.08	1.34	–	4
Mirny	66°33'S	93°01'E	820	45	1.23	1.22	–	5
Little America	78°10'S	162°13'W	0	0	1.08	1.79	–	6
Ross Ice Shelf C7	78°58'S	176°00'W	0	70	0.94	1.13	–	7
Ross Ice Shelf Base Camp	82°28'S	166°00'W	0	450	1.19	1.13	–	7
Ross Ice Shelf site E, F, G	–	–	–	650	1.48	3.94	–	7
Dome C	74°42'S	124°04'E	3240	910	1.24	1.69	1.69	8
Dome C	74°42'S	124°04'E	3240	910	1.04	2.10	1.73	9
Dome C	74°42'S	124°04'E	3240	910	1.10	2.05	1.64	10
James Ross Island Na > 0.25 mg/kg	64°13'S	57°38'W	1660	24	–	0.83	1.11	1
James Ross Island Na < 0.25 mg/kg	64°13'S	57°38'W	1660	24	–	3.73	5.35	1
South Pole	90°S	–	2880	1270	1.20	2.75	2.47	11
Vostok	78°28'S	106°48'E	3488	1300	–	1.88	–	2
South Pole	90°S	–	2880	1270	–	2.11	–	12
Riiser Larsenisen ice shelf	72°30'S	15°00'W	–	1	1.16	–	1.10	13
Riiser Larsenisen ice shelf	72°30'S	15°00'W	–	60	1.28	–	1.34	13
Riiser Larsenisen ice shelf	72°30'S	15°00'W	–	120	1.11	–	1.41	13
Byrd Station (0–1000 m)	79°59'S	120°01'W	1530	700	1.11	2.78	–	14
Crescent Scarp (peninsula)	69°42'S	66°25'W	1500	90	–	0.95	0.98	15
Law Dome	66°30'S	113°00'E	–	115	0.77	–	–	16
Vostok (0–325 m)	78°28'S	106°48'E	3488	1300	1.03	1.56	1.90	17
South Pole (1959–69)	90°S	–	2880	1270	–	2.11	–	18
South Pole (last 1000 y)	90°S	–	2880	1270	–	1.89	–	18
Dome C (238–360 m)	74°42'S	124°04'E	3240	910	–	1.06	–	19
Dome C (360–454 m)	74°42'S	124°04'E	3240	910	0.99	0.89	–	19
Dumont D'Urville	66°42'S	140°00'E	41	–	1.05	1.29	1.01	20
South Pole (1955–88)	90°S	–	2880	1270	2.10	2.75	9.98	21
Byrd Station	79°59'S	120°01'W	1530	700	1.17	–	–	22
Mc Carthy Ridge	74°32'S	162°56'E	700	40	0.95	1.86	2.16	–
Styx Glacier	73°51'S	163°41'E	1700	50	0.94	1.63	2.95	

The purpose of this paper is to discuss the non-sea spray effects on two series of snow samples obtained from pits opened in Northern Victoria Land (Antarctica).

EXPERIMENTAL

Sampling and analysis

In mid December 1990 two snow pits, 2 m deep, were dug at Styx Glacier (1700 m a.s.l., and 50 Km distance from the sea) and at Mc Carthy Ridge (700 m a.s.l., 40 Km distance from the sea). Two series of progressively numbered vials, one 16 × 100 mm and another 35 × 100 mm, were inserted in the walls of each pit. The small diameter vials were divided into even and odd numbers. The snow of the even numbered vials were analysed for H_2O_2 while the odd ones for only anions. The results of these analysis have already been reported[23]. The snow of the large diameter vials was analysed for H_2O_2, anions and cations. These analysis provide a more complete characterization of a single snow layer.

The hydrogen peroxide was determined by flow analysis using p-hydroxyphenylacetic acid and peroxidase and fluorescence intensity measurement.

Anions were determined by Ion-chromatography using a Dionex AS5A column for fluorides and organic acids (acetic, formic and methanesulphonic (MSA)) and a Dionex AS4A for inorganic anions. The eluents used were, respectively, 3.3 mM borax and 1.2 mM $NaHCO_3$ + 1.3 mM Na_2CO_3.[24] The instrument was equipped with a conductivity detector with suppression.

Cations were determined by ion chromatography using a Dionex CS10 column, eluted with 20 mM HCI + 0.5 mM diaminopropionic acid[24] and a conductivity detector with suppression.

Samples were introduced into the chromatograph immediately after melting and filtration on 0.45 μm membrane so to avoid eluent effects on eventually present suspended solid matter. In fact, as described by De Angelis *et al.*[2], only the Na of marine origin is present in the melting. The Na of alluminosilicates is completely solubilized passing through the chromatographic column[25]. In absence of a preliminary filtration, Na of marine origin was recognised from crustal Na measuring the Al content and considering the ratio of mean concentrations in the earth crust[26].

Further indications on the sampling mode and laboratory activity have been described elsewhere[23].

RESULTS AND DISCUSSION

As previously observed[23], the H_2O_2 concentration show a cyclic behaviour with spring maximum, which allows the recovery of annual layers. The H_2O_2 concentration as a function of samples depth for both pit are reported in the continuos curve of Figure 1. The depositions in the Styx Glacier pit, dated according to the H_2O_2 records, correspond to the period 1986–1990, with a mean resolution of 10 samples per year. On the other hand, the Mc Carthy Ridge pit covers the period 1987–1990 with a mean resolution of 15 samples per year. This dating is confirmed by a plot obtained from the cross comparison of the concentration of H_2O_2, $nssSO_4$, and MSA[27]. A plot of Cl/Na versus depth confirms the time series for Styx Glacier, whereas, for the lower sampling place, the identification of the time series is not possible, as the curve does not show a well defined trend.

The Na concentration profiles (bar histograms in Figure 1), show very high values primarily in winter samples. Profiles obtained for different ions which are the main constituents of sea water (i.e. Cl, Mg, Ca, K and total sulfate), evidence high values of concentration at the same depth. Besides, considering for these elements the concentration ratio X/Na with increasing Na concentrations, the data points follow the line relative to the X/Na ratio of sea water. Assuming that the samples with high Na content have been deposited during salt storms, it is evident that the samples collected must be treated differently, and two groups with high Na content (H-Na) and low Na content (L-Na) must be determined for each pit. The frequency histograms of Na concentration of both sampling places were constructed in order to identify the threshold between the two groups of samples. The analysis of distribution show that most samples (78% and 83% for Styx Glacier and Mc Carthy Ridge, respectively) are grouped in contiguous classes of frequency in the range of low Na concentrations. The least samples are distributed in frequency classes well separated from the previous samples in the range of high Na concentrations. On the basis of these considerations, the threshold limits were established to be 0.25 and 1.0 mg Na/Kg for Styx Glacier and Mc Carthy

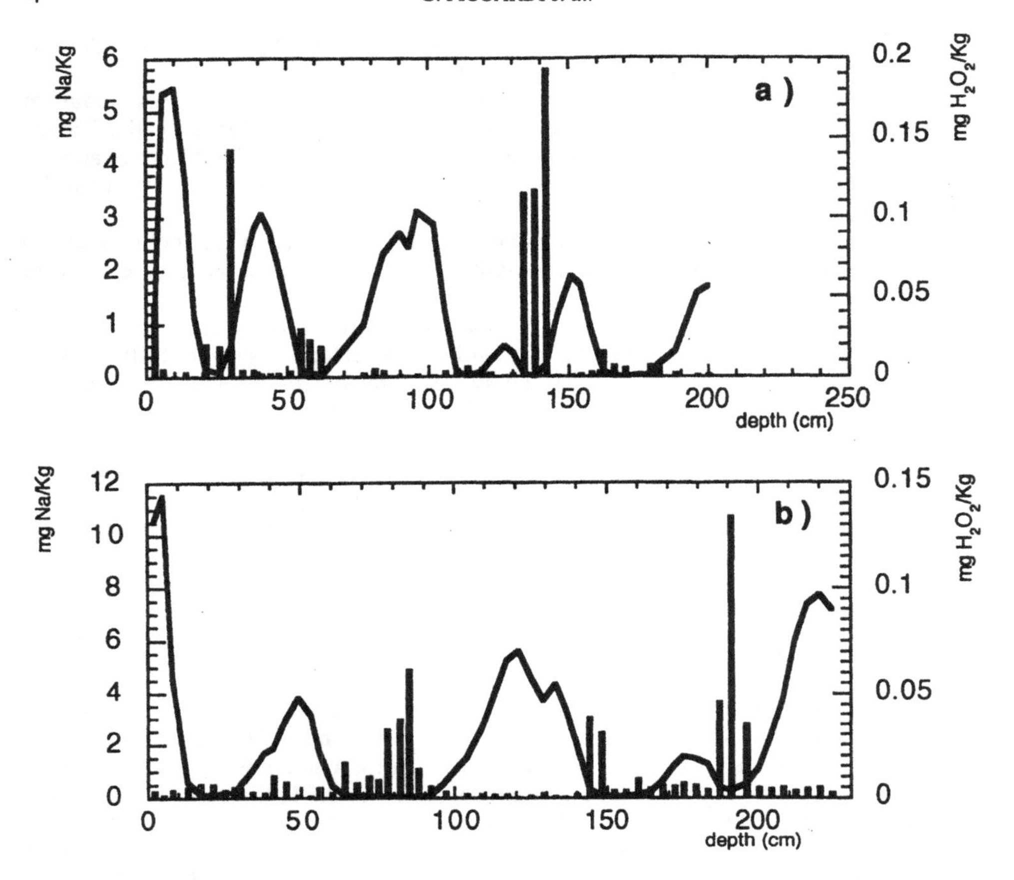

Figure 1 Time series of H_2O_2 (line) and Na (bars) in a) Styx Glacier, b) Mc Carthy Ridge samples.

Ridge, respectively. The same threshold limits were obtained in the analysis of clusters of the Euclidean distances among samples in space defined by the Na, Cl, Mg, Ca, K, and sulfates variables[28].

A further classification was introduced, on the basis of H_2O_2 concentration, to follow the time series of layer sequences. A biased value of 10 µg H_2O_2/Kg was used to distinguish the summer from winter samples. A so small value for H_2O_2 concentration was chosen to include the small maximum relative to the Antarctic summer of 1987–1988. This annual cycle, whose existence could be questioned on the basis of the H_2O_2 profile only (Figure 1), is in fact confirmed by the other measures considered for dating (nss-SO_4, MSA, and Cl/Na). After this classification, the samples of the H-Na group are all winter samples for the Mc Carthy Ridge sampling place, whereas in the Styx Glacier, two summer samples are present. Winter and summer samples are present in the L-Na group for both sampling places, with a prevailing presence of summer samples.

Once the possibility to separate the samples in three groups has been established (L-Na/s = summer, L-Na/w and H-Na/w = winter), the H-Na/s practically does not exist. The mean concentration of Na, Mg, K, Ca and sulfates for each group was calculated and, from these data, the E.F. were obtained with respect to Na (Table 2). It can be noted that the E.F. increases from H-Na/w to L-Na/s and that Cl and Mg have a different behaviour from K and Ca. For the first two ions small variations of E.F. are observed (values around 1), very close to the sea salt ratio. On the contrary, for K and Ca, large variations of E.F. are observed, reaching the value of 15 for Ca at Styx Glacier. In Table 2 (last column) is also reported the value of mean Na concentration used for calculating the single E.F. values; this allows to observe an inverse relationship between E.F relative to K and to Ca and the mean Na concentrations. This fact could lead to hypothesise a constant contribution to E.F. of nssK and nssCa which became less important as the contribution of sea K and Ca increases.

The values of E.F. obtained from our measurements as a mean of all samples of each sampling place are also reported in Table 1. As we can see, the value of E.F for Mg is about 1, whereas those of K and Ca are about the double. Therefore, the E.F. calculated from the mean of all samples, as usually reported, bring only a modest contribution to the knowledge of the phenomenon.

On the other hand, as indicated in Table 1, different authors report values always close to 1 for E.F. relative to Mg and values often greater than 1 for K and Ca, indicating that also in different regions of Anctartica it exists a different behaviour of Mg, primarily of marine origin, in comparison to K and Ca, which would have other origin in addition to sea salt contribution. An example of separation based on Na content has been attempted by Aristarain *et al.*[1] who separated the samples collected at Ross Island in two series. The E.F. values of K and Ca at L-Na resulted higher than those of H-Na. Analogous conclusions were drown by Warburton *et al.*[29] from the analysis of Ross Ice Shelf snow. Tuncel *et al.*[3] separated aerosol samples collected at Amundsen-Scott base on seasonal basis. The E.F. values of Mg, K and Ca, calculated for winter samples resulted 1.23, 1.56 and 1.53, respectively, whereas, for the summer samples, values of 1.53, 2.58, and 3.03 were found. The contribution of the marine source to the different variables was investigated, through the analysis of linear regression versus Na used as tracer of the sea-salt source, on groups separated for each sampling place (four plots = 2 sampling places × 2 groups, H-Na and L-Na). Each plot reports the value of the ratio corresponding to bulk sea water (dashed line) and the regression line (continuous line) when the model significance is greater than 99%. As it concerns the H-Na samples relative to Styx Glacier, the only two summer samples show a behaviour very different from that of the winter samples. For this reason the regression lines relative to H-Na are all calculated on winter samples.

Table 2 Enrichment factors (E.F.) of Cl, Mg, K, Ca and sulfates in samples of Styx Glacier and Mc Carthy Ridge.

		Cl	*Mg*	*K*	*Ca*	*SO_4*	*Mean Na concentration mg/Kg*
Styx Glacier	H-Na/w	0.87	0.80	1.20	1.49	0.93	1.858
	L-Na/w	0.97	0.83	3.43	5.58	1.23	0.178
	L-Na/s	1.18	1.07	4.22	15.31	3.85	0.077
Mc Carthy Ridge	H-Na/w	1.03	0.88	1.13	1.33	1.04	3.588
	L-Na/w	0.93	0.90	3.11	3.49	1.55	0.471
	L-Na/s	1.00	0.90	3.58	4.18	2.79	0.295

Chlorides

For the H-Na samples in both sampling places a good correlation between Cl and Na was observed (Figures 2a and 2c). The regression of Cl over Na, relative to the sea salt (theoretical regression) is basically coincident to the regression calculated for the samples of the lower station, whereas, for the higher station, the data points are mostly lower than that of the sea salt. For the L-Na samples of the Styx Glacier (Figure 2b) the sample regression crosses the theoretical regression near the value of 0.14 mg Na/Kg, because the samples with a higher Na concentration, mostly winter samples, have Cl concentrations lower than that of the sea salt. On the contrary the summer samples, characterised by lower Na content, show a Cl/Na ratio higher than the sea salt. For this reason the Cl/Na ratio can give a valuable indication of the season alternance. In the

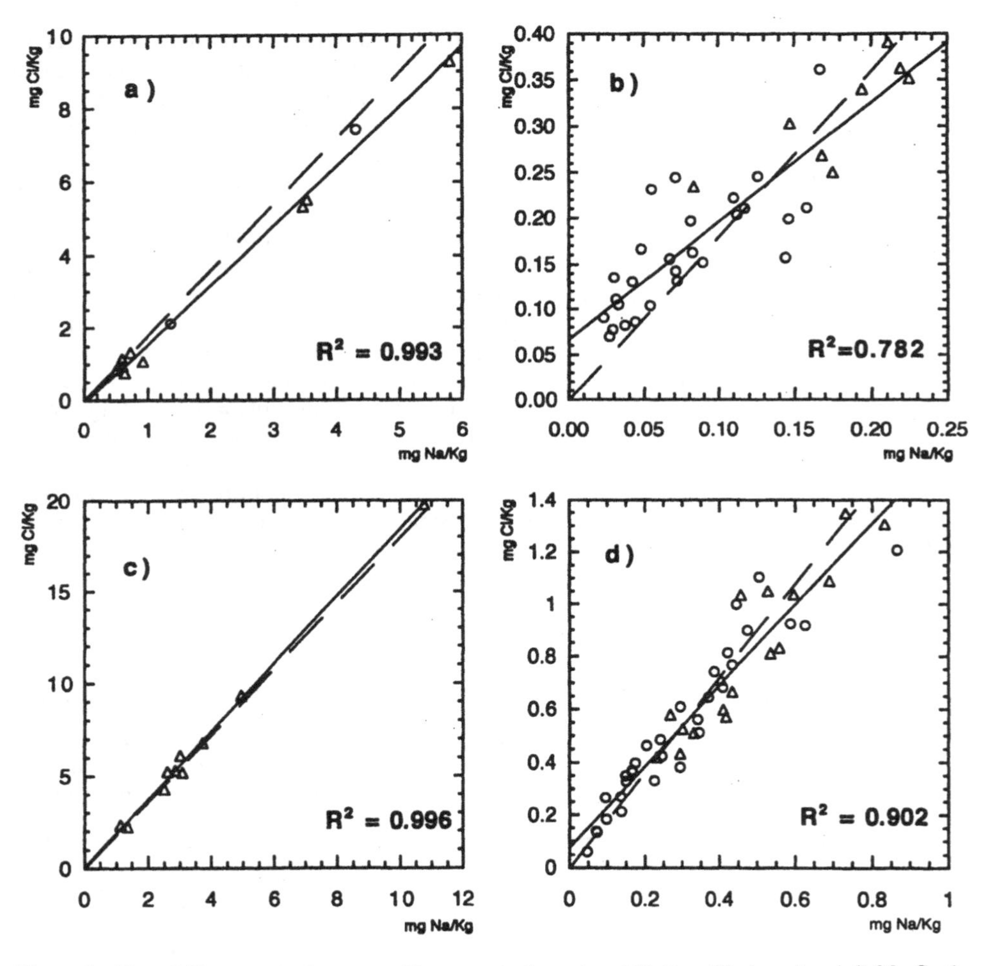

Figure 2 Plot of Cl concentration versus Na concentration; a) and b) Styx Glacier, c) and d) Mc Carthy Ridge; a) and c) H-Na, b) and d) L-Na; o summer samples, Δ winter samples; ——— regression line, — — sea water composition.

lower station (Figure 2d) the phenomenon occur with weaker intensity and the Cl/Na ratio is not reliable marker of season alternance. Legrand and Delmas[19] noticed that the Cl/Na ratio is generally very close to that of bulk sea water near the coast and begins to increase at the edge of the Antarctic plateau. Inasmuch, this ratio has seasonal variations with evident summer maxima at South Pole[12,21].

Magnesium

The magnesium ion show a behaviour similar to that of Cl (Figure 3). The H-Na samples of the higher station are primarily depleted of Mg. Only the two summer samples have different behaviour in comparison to the winter samples. As for L-Na samples, the regression line, always significative, intercept the theoretical regression, as the winter

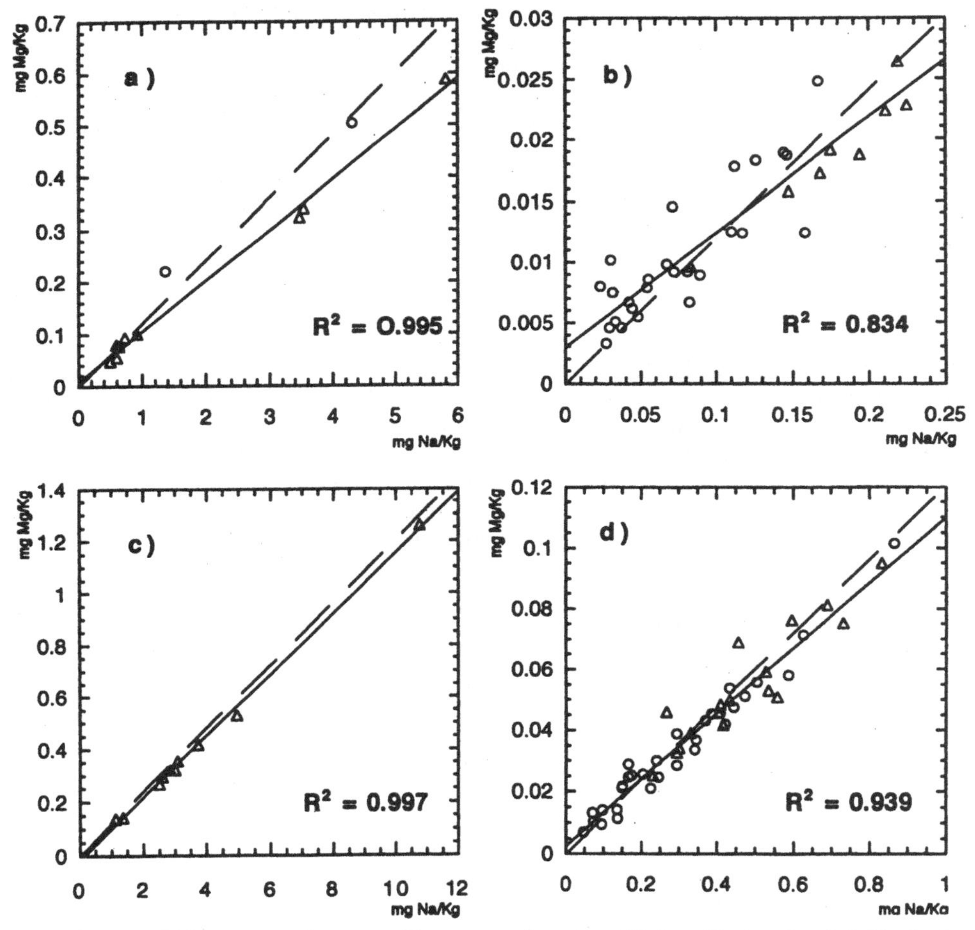

Figure 3 Plot of Mg, concentration versus Na concentration; a) and b) Styx Glacier, c) and d) Mc Carthy Ridge; a) and c) H-Na, b) and d) L-Na; o summer samples, Δ winter samples; ——— regression line, — — sea water composition.

samples are lower in Mg and summer samples are primarily enriched. Also for this element it is possible to establish a seasonal alternance from the Mg/Na ratio.

The samples from Mc Carthy Ridge, i.e. the lower station (Figure 3c), are practically aligned to the theoretical regression. The modest deviation of the values of Cl and Mg concentration from the theoretical regression evidences that, for this station, the fractionating process is only partially active, and that the transport processes remain conservative in the areas along the coast. The good correlation between Mg and Cl ($R^2 = 0.925$) found by Mulvaney *et al.*[30] in samples from Filibusen, lead the authors to hypothesise that the two species were related in the aerosol even as $MgCl_2$. A non negligible percentage (5%) of $MgCl_2$ particles was observed by Artaxo *et al.*[31] in an aerosol of Brazilian Antarctic Station at King George Island.

The use of sodium as an indicator of marine contribution seems to be justified, in this case, by the good correlation of Cl and Mg with Na (Figures 2 and 3) and of Cl with Mg (Styx Glacier $R^2 = 0.73$, Mc Carthy Ridge $R^2 = 0.89$), which lead to hypothesise that all three elements have the same origin, being in the same ratio as in sea salt. The plots of Figures 2 and 3 confirm the data in Table 2 where the E.F relative to Cl and Mg are close to 1. As indicated in Table 1, where the E.F. values are, also in this case, close to unity, the phenomenon is not restricted to our sampling area confirming, unlike K and Ca, the marine origin of Mg. The concentration of Mg in South Pole aerosol is controlled by marine aerosol all year along[3]. The lack of a crustal contribution for Mg is revealed not only in the soluble component of the snow melt, but also in the suspended solid fraction. The data of Boutron[9–11,32] relative to samples collected at Dome C and in South Pole, performed after a solubilization of aluminosilicates with HF, show, also in this case, the lack of a Mg excess with respect to marine Mg.

Potassium

In samples classified as H-Na of both sampling places (Figures 4a and 4c) a very good linear regression between K and Na is observed and the slopes are similar to that of the sea salt as the regression lines are parallel to the theoretical regression. The intercepts of these lines are significative, different from zero, and corresponding to 16 and 10 μg K/Kg for Styx Glacier and Mc Carthy Ridge, respectively. In the higher station one of the two samples classified as summer samples shows a K/Na ratio much greater with respect to that of winter samples (Figure 4a). The L-Na samples of Mc Carthy Ridge (Figure 4d) show again a parallelism with the theoretical regression albeit with a low regression coefficient. In this case also all the samples are richer in K and the intercept indicate a contribution of about 20 μg K/Kg. For the higher station it is not possible to establish any regression (Figure 4b) and all samples have a K concentration higher than the sea salt and the median is about 11 μg/kg. Therefore, a constant background contribution exists for K in both station, that is independent of marine contribution.

Calcium

Calcium behaves mostly like potassium. For this element also the winter samples of the H-Na group of both sampling station lies on lines parallel to the theoretical regression with a significative intercept of 40 μg Ca/Kg for both stations (Figures 5a and 5c). As observed for Mg at Styx Glacier, both samples classified as summer samples have higher Ca concentration in comparison with the corresponding winter samples (Figure 5a).

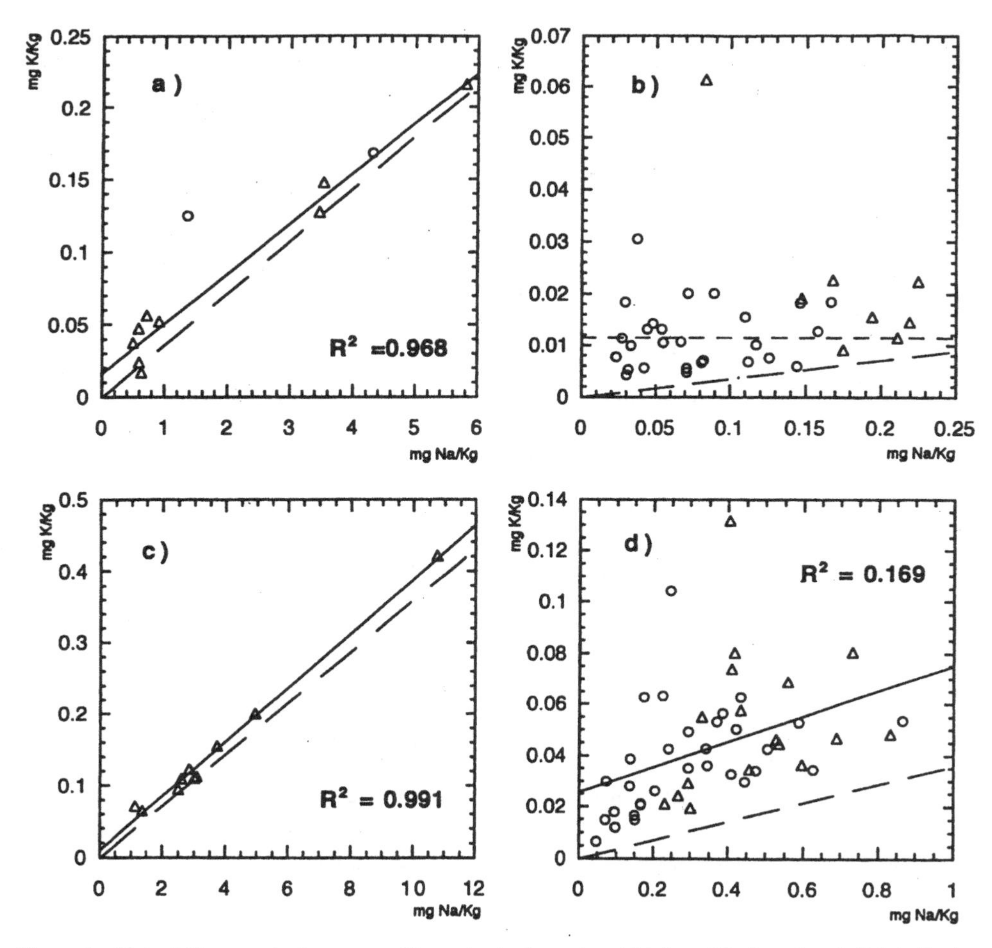

Figure 4 Plot of K concentration versus Na concentration; a) and b) Styx Glacier, c) and d) Mc Carthy Ridge; a) and c) H-Na, b) and d) L-Na; o summer samples, Δ winter samples; ——— regression line, — · — sea water composition, – – – – median concentration.

All the L-Na samples (Figures 5b and 5d) show a Ca excess in comparison with the sea salt contribution and, also in this case, the L-Na samples of Styx Glacier have a median concentration similar to that revealed by the intercept of the H-Na samples. The K/Na and Ca/Na time series do not evidence any appreciable seasonal trend, as the background noise is high. On the other hand, the median values of the L-Na samples are comparable both for Ca and K with the values of the intercept of the H-Na samples, suggesting that the contribution of marine spray is added to contributions of different origin, homogeneously distributed and particularly evident when the weather conditions do not facilitate the sea salt input.

A different behaviour between Mg from one side, and Ca and K from the other is evidenced in Table 2. The same difference has been evidenced in different regions of Antarctica as reported in Table 1. The presence of an excess of Ca and K

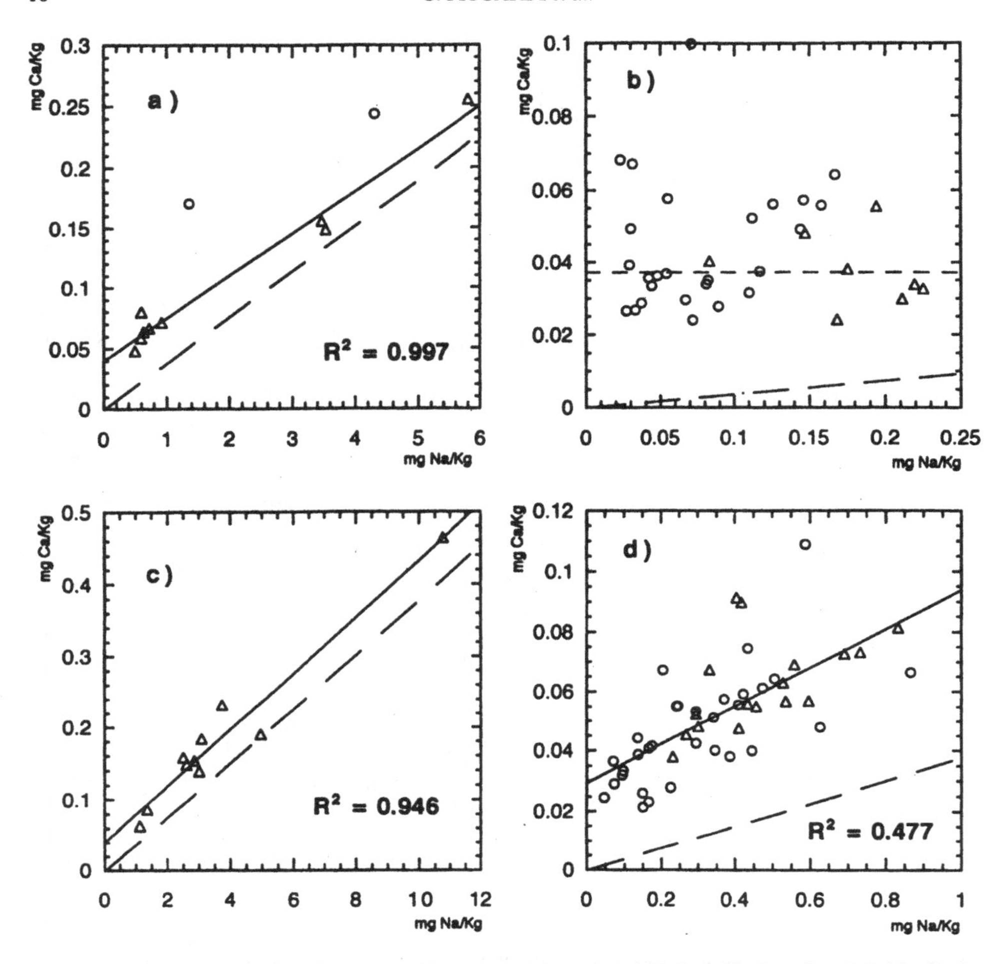

Figure 5 Plot of Ca concentration versus Na concentration; a) and b) Styx Glacier, c) and d) Mc Carthy Ridge; a) and c) H-Na, b) and d) L-Na; o summer samples, Δ winter samples; ——— regression line, — — sea water composition, – – – – median concentration.

contents compared to sea salt (Na and Mg) has also been found in aerosols of South Pole[3,33], Syowa Station[34], King Geoge Island, Antarctic Peninsula[31] and East Antarctic Plateau[35].

The Ca excess is ascribed to a contribution of crustal[13] or marine biogenic (coccoliths)[36] $CaCO_3$. The lack of carbonates in ice observed by Legrand *et al.*[37] lead to hypothesise that the nssCa is linked with both $nssSO_4$ and NO_3, the presence of $CaSO_4$ cristals in Antarctic aerosol has been evidenced by scanning electron microscopy and PIXE[31,36]. Also for K the crustal origin has been proposed[7]. A few authors support the idea that the excess of various substances in aerosols is the result of a chemical fractionation during the production of atmospheric sea salt particles by bursting bubbles[38,39].

Sulfates

The regressions between sulfate and sodium, for the H-Na group, are similar to those of Cl and Mg (Figures 6a and c). The correlation coefficients are high and the slopes close to those of the theoretical regression and for these samples, essentially winter samples, the marine spray is the main source of sulfates. Also in this case the two summer samples of Styx Glacier have a different behaviour compared with the winter samples (Figure 6a).

On the contrary for the L-Na group (Figures 6b and 6d), in which prevalently summer samples are included, the behaviour is similar to that of K and Ca. The low significance of the correlations over Na (95% probability level) and concentrations higher than the sea salt ratio indicate a contribution of different origin.

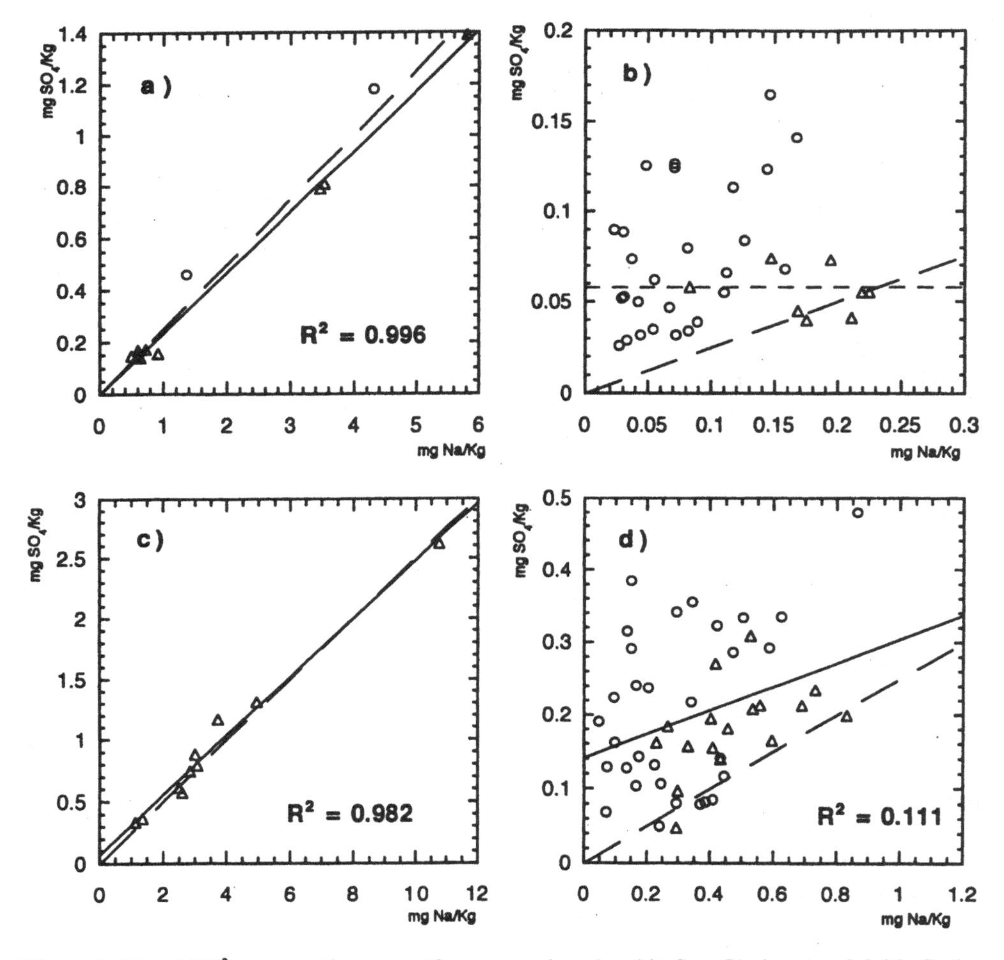

Figure 6 Plot of SO_4^{2-} concentration versus Na concentration; a) and b) Styx Glacier, c) and d) Mc Carthy Ridge; a) and c) H-Na, b) and d) L-Na; o summer samples, Δ winter samples; ——— regression line, — — sea water composition, - - - - median concentration.

From the median concentration values used for Table 2 it is possible to estimate that, in L-Na/s samples, 26% of the Styx Glacier and 36% of the Mc Carthy Ridge sulfates are of sea origin. For both stations this value is 100 % in H-Na/w samples.

The decrease of the marine contribution with the distance from the coast has been evidenced by Mulvaney *et al.*[30] determining the concentrations of Na and sulfates along a transect of the Fimbul Ice Shelf in Dronning Maud Land. Marine sulfates of shallow samples collected during January 1990 have a mean value of 20% of the total sulfates. This percentage changes along the transect from 45% near the coast to 10% at 120 Km of distance from the sea[30]. Previous papers evidenced that variations in the chemistry of snow are not a simple function of distance from the sea[7] but that the critical parameter for the transport of bulk sea-salt aerosol is elevation rather than distance from the coast[40]. This reduction of percentage of marine sulfates may be ascribed to the slow oxidation of SO_2, but also to a possible contribution of $nssSO_4$ from long range sources, or to a shorter residence time of the sea-salt particles[30]. Whitlow *et al.*[21] found that the marine contribution is only 5% at the South Pole.

In the southern hemisphere the biogenic source of sulphur became prevalent because the anthropogenic contribution is practically negligible[41]. In fact, considering the analogies between the time series of sulfates of the two stations[23], the source of $nssSO_4$ may be supposed to be the biogenic sulfate produced in the oxidation of dimethylsulfide (DMS). Strong covariations of MSA and $nssSO_4$, determined in aerosols, have been observed in several remote marine stations[42,43], but evident seasonal cycles are found above all in high latitude stations: Shemya Island (52°N)[42], Cape Grim (40°S), where a good relation was found among DMS, MSA and $nssSO_4$ determined in the atmosphere[44–46], Amsterdam Island (38°S)[47] and Mawson, East Antarctica (67°S)[48]. The good correlation among the above mentioned components strongly indicates their common source.

The sulphur cycle in the ocean-atmosphere system is influenced by climatic variability and produces changes in the $nssSO_4$/MSA concentration ratio.

In addition to the temperature effect on branching ratio of the oxidation of DMS (this effect was observed in laboratory experiments by Hynes[49] and supported by Berresheim[50] to explain the variation of the MSA/$nssSO_4$ ratio with latitude, even if the phenomenon was not always observed[44,48]), the wind velocity plays an important role both on oceanic emissions of air/sea exchanging rates[51] and on the abundance of sulfur dioxide and fine particle sulfate aerosols and the resulting size distribution of the marine aerosols[52].

A preliminary evaluation of errors associated with the determination of nss-SO_4 concentrations is necessary to usefully utilise these values. In fact, as the analytical error associated with the direct determination of Na and SO_4 is less than 5%, (3% using our methodology[24]), it is necessary to remember that the concentration of $nssSO_4$ results from the difference of two concentration values. Therefore, a calculation procedure[53] has been applied for the relative analytical precision, and the discussion is restricted only to values with a relative error less than 10%. So, all H-Na samples and most of the L-Na/w were discarded, and the drawn conclusions may be considered as valid for summer samples only.

The plots reported in Figure 7 were constructed using these data to evaluate the contribution of $nssSO_4$ (possibly of different origin) in comparison with the MSA concentration (of sole biogenic origin).

For the Mc Carthy Ridge samples (Figure 7b), a good linear relation was found between $nssSO_4$ and MSA concentrations following the equation:

$[nssSO_4] = 1.89\,[MSA] + 50.8$; $N = 25$, $R^2 = 0.75$

std. error	0.23	16.4
sig. level	0.000	0.005

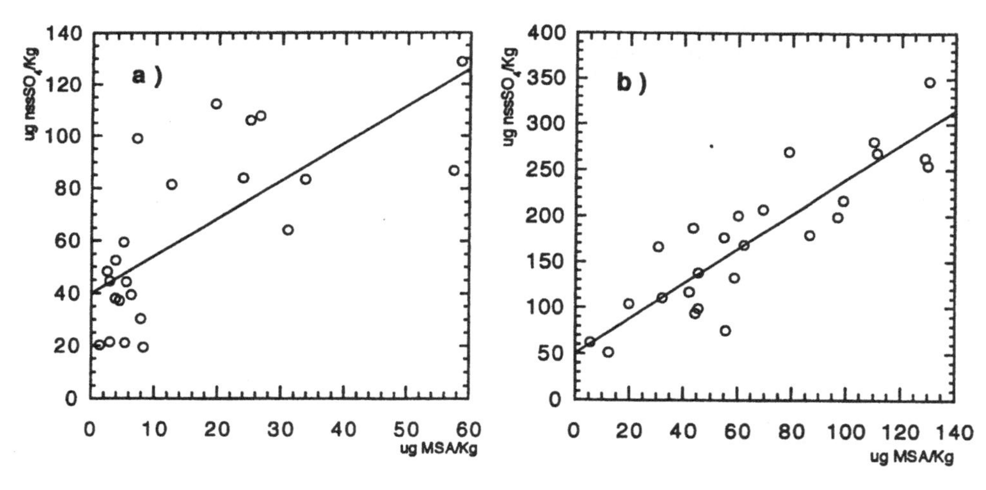

Figure 7 Plot of $nssSO_4$ concentration versus Na concentration; a) styx Glacier, b) Mc Carthy Ridge; ——— regression line.

The intercept is significant different from zero (95% probability level) and the slope is comparable with the mean values $nssSO_4$/MSA reported for ice cores and fresh snow from other coastal areas of Antarctica. For example, the following values are reported: 2.56 for an ice core of Law Dome east Antarctica[16], 2.5–10 for fresh summer snow of Dumont D'Urville[20] and 2.17–3.13 for ice cores of Antarctic peninsula[55]. For the Styx Glacier station, the linear association degree of $nssSO_4$ with MSA is much lower even though highly significant (99%). For this station, significant correlations were found also for NO_3 and nssCa, as reported in Table 3 in which, besides the correlation coefficients, the relative significant are indicated.

The variance of $nssSO_4$ is almost completely explained by the multiple regression model with MSA and NO_3 (Table 4a) or by the regression with MSA and nssCa (Table 4b). If all three variables (MSA, NO_3 and nssCa) are inserted in the regression model (Table 4c), no improvement is observed. A highly significant correlation between NO_3 and $nssSO_4$ is shown in Table 3, and the common variance between $nssSO_4$ and nssCa is already included in the model a, in Table 4. These considerations lead to conclude that in this station another source of $nssSO_4$ appears[56], in addition to that produced by DMS oxidation, which is correlated with NO_3 and/or with nssCa. A long range continental contribution of $nssSO_4$ has been recognised by Savoie *et al.*[57] at Mawson, from the relation between $nssSO_4$ and NO_3 and from the covariance of the latter with ^{210}Pb. This hypothesis can be extended also to the Styx Glacier station and an analogous behaviour is observed also with Mc Carthy Ridge samples (Table 4 d–f).

A confirmation of the different contributions to $nssSO_4$ in the two stations arises from the comparison of the mean $nssSO_4$/MSA ratio distributions: 3.27 (s = 2.00) for Mc Carthy Ridge, and 7.53 (s = 5.21) for Styx Glacier.

The increase of the $nssSO_4$/MSA ratio with altitude was reported also by Legrand and Saigne[57] comparing the values measured in ice cores sampled in both coastal and inland areas of Antacrtica. The values ranged from 2.27–5.26 for an ice core of D10, Terre Adelie[58] to values high as 10–33 measured on ice cores of the inland[59]. According to

Table 3 Correlation coefficients and significance levels in samples of Styx Glacier and Mc Carthy Ridge.

	$nssSO_4$	MSA	NO_3	$nssCa$
Styx Glacier				
$nssSO_4$	1.000	0.715	0.616	0.504
sig. level	–	0.000	0.002	0.017
MSA	0.715	1.000	0.083	0.168
sig. level	0.000	–	0.708	0.454
NO_3	0.616	0.083	1.000	0.601
sig. level	0.002	0.708	–	0.003
nssCa	0.504	0.168	0.601	1.000
sig. level	0.017	0.454	0.003	–
Mc Carthy Ridge				
$nssSO_4$	1.000	0.871	0.146	–0.102
sig. level	–	0.000	0.497	0.635
MSA	0.871	1.000	–0.181	–0.346
sig. level	0.000	–	0.398	0.098
NO_3	0.146	–0.181	1.000	0.351
sig. level	0.497	0.398	–	0.092
nssCa	–0.102	–0.346	0.351	1.000
sig. level	0.635	0.098	0.092	–

Shao-Meng LI *et al.*[60] the $nssSO_4$/MSA ratio increases with altitude also in the Arctic zone, with values ranging from 5.88 in Alert (Canada) to 16.67 in Dye3 (Greenland 2480 m a.s.l.) up to 20 at Summit (Greenland, 3200 m a.s..l).

Different hypotheses have been proposed to explain the variation of this ratio with altitude. Legrand[58] proposes a difference in origin of the air masses that transport $nssSO_4$ and MSA: for the coastal areas the air masses would origin by the closely ocean, whereas for the inland areas the air masses would came from the temperate zone. For these reasons the variation of the $nssSO_4$/MSA ratio depends upon the latitude at which the air masses originate. Other possible explanations involve: a) a different distribution of $nssSO_4$ and MSA in aerosols, with a subsequent modification of their deposition processes; b) different oxidation mechanisms of DMS with altitude; c) the existence of different contributions of $nssSO_4$ not originated by DMS oxidation.

As the two considered sampling stations are located at the same distance from the coast, the air masses active in both locations are presumably not different, and a different contribution of sulfates is responsible for the modified $nssSO_4$/MSA ratio[56].

Some unclarified aspects still exist in the cycle linking DMS, MSA and $nssSO_4$ albeit this cycle could have an important role in global climate changes as postulated by Charlson[61]. The use of MSA as an indicator of the biogenic source contribution is complicated by the variability, in space and time, of the $nssSO_4$/MSA ratios, that can be very different even in stations relatively close, as demonstrated in the Terra Nova Bay area.

Table 4 Multiple regression models in samples of Styx Glacier and Mc Carthy Ridge.

Styx Glacier				
mod. a) $[nssSO_4] = 1.34[MSA] + 0.29[NO_3] + 14.6$				
std. error	0.19	0.05	6.0	
sig. level	0.000	0.000	0.025	
$R^2 = 0.806$				
mod. b) $[nssSO_4] = 1.26[MSA] + 0.75[nssCa] + 8.53$				
std. error	0.26	0.26	12.1	
sig. level	0.000	0.009	0.488	
$R^2 = 0.626$				
mod. c) $[nssSO_4] = 1.30[MSA] + 0.25[NO_3] + 0.17[nssCa] + 10.4$				
std. error	0.20	0.06	0.24	9.03
sig. level	0.000	0.001	0.486	0.263
$R^2 = 0.791$				
Mc Carthy Ridge				
mod. d) $[nssSO_4] = 2.02[MSA] + 0.47[NO_3] + 9.53$				
std. error	0.18	0.13	17.2	
sig. level	0.000	0.001	0.585	
$R^2 = 0.841$				
mod. e) $[nssSO_4] = 2.06[MSA] + 1.30[nssCa] - 5.35$				
std. error	0.22	0.59	29.7	
sig. level	0.000	0.038	0.859	
$R^2 = 0.786$				
mod. f) $[nssSO_4] = 2.10[MSA] + 0.41[NO_3] + 0.79[nssCa] - 19.0$				
std. error	0.19	0.13	0.52	25.3
sig. level	0.000	0.005	0.148	0.460
$R^2 = 0.850$				

CONCLUSIONS

From the composition of snow samples collected in the area of Terra Nova Bay the different contributions of main elements may be individualized. Since the marine contribution changes with the climatic conditions, the E.F. obtained from mean concentrations, as usually reported, does not reveal the true effect of different sources.

The grouping samples on the basis of sea-salt content or seasonality exhibit a better information on the enrichment processes. Regression plots of different components versus sodium, considered an indicator of marine contribution, individualize quantitatively the non-sea-salt effect. So, while for Na, Cl and Mg only a marine contribution is observed, a constant excess of non sea spray K and Ca suggests the existence of a nearly uniform background aerosol over the investigated area.

The dominant source of $nssSO_4$ is the DMS oxidation but a very significant fraction is associated with NO_3 or nssCa. Significant correlations were found in the multiple variable regression analysis of $nssSO_4$, MSA, NO_3 and nssCa. The non-DMS fraction may be attributed to long range transport phenomena.

References

1. A. J. Aristarain, R. J. Delmas and M. Briat, *J.Geophys.Res.*, **87**, 11004–11012 (1982).
2. M. De Angelis, M. Legrand, J. R. Petit, N. I. Barkov, Ye. S. Korotkevitch and V. M. Kotlyakov, *J. Atmos. Chem.*, **1**, 215–239 (1984)
3. G. Tuncel, N. K. Aras and W.H. Zoller, *J. Geophys. Res.*, **94**, 13025– 13038 (1989).
4. M. Murozumi, T. J. Chow and C. Patterson, *Geochim. Cosmochim.Acta*, **33**, 1247–1294 (1969).
5. C. Boutron, M. Echevin and C. Lorius, *Geochim. Cosmochim. Acta*, **36**, 1029–1041 (1972).
6. C. C. Langway jr, M. M Herron and J. H. Cragin, *U.S. Cold Regions Research and Engineering Laboratory Research Report*, p. 77, (1974).
7. J. A. Warburton and G. O. Linkletter, *J. Glac.*, **20**, 149–162 (1978).
8. C. Boutron and C. Lorius, *Nature*, **277**, 551–554 (1979).
9. C. Boutron, *J. Geophys. Res.*, **85**, 7426–7432 (1980).
10. C. Boutron and S. Martin, *J. Geophys. Res.*, **85**, 5631–5638 (1980).
11. C. Boutron, *Atmos. Environ.*, **16**, 2451–2459 (1982).
12. M. R. Legrand and R. J. Delmas, *Atmos. Environ.*, **18**, 1867–1874 (1984).
13 Y. Gjessing, *Atmos. Environ.*, **18**, 825–830 (1984).
14. J. M. Palais and M. Legrand, *J. Geophys. Res.*, **90**, 1143–1154 (1985).
15. A. L. Dick and D. A. Peel, *Ann. Glac.*, **7**, 12–19 (1985).
16. J. P. Ivey, D. M. Davies, V. Morgan and G. P. Ayers, *Tellus*, **38B**, 375–379 (1986).
17. M. R. Legrand, C. Lorius, N. I. Barkov and V. N. Petrov, *Atmos. Environ.*, **22**, 317–331 (1988).
18. S. Kirchner and R. J. Delmas, *Ann. Glaciol.*, **10**, 1–5 (1988).
19. M. R. Legrand and R. J. Delmas, *J. Geophys. Res.*, **93**, 7153–7168 (1988).
20. F. Maupetit and R. J.Delmas, *J. Atmos. Chem.*, **14**, 31–42 (1992).
21. S. Whitlow, P. A. Mayewski and J. E. Dibb, *Atmos. Envir.*, **26A**, 2045–2054 (1992).
22. C. C. Langway, K. Osada, H. B. Clausen, C. U. Hammer, H. Shoji and A. Mitani, *Tellus*, **46B**, 40–51 (1994).
23. G. Piccardi, R. Udisti and F. Casella, *Intern. J. Environ. Anal. Chem.*, **55**, 219–234 (1994).
24. R. Udisti, S. Bellandi and G. Piccardi, *Fresenius J. Anal. Chem.* **349**, 289–293 (1994).
25. M. Legrand, *J. de Physique*, **48**, C1 77–C1 86 (1987).
26. M. Legrand, C. Lorius, N.I. Barkov and V. N. Petrov, *Atmos. Environ.*, **22**, 317–331 (1988).
27. R. Udisti, *Intern. J. Environ. Anal. Chem.*, **63**, (1996).
28. F. Casella, R. Udisti and G. Piccardi, *Environmetrics*, (submitted).
29. J. A. Warburton, C. R. Cornish, J. V. Molenar, M. S. Owens and L. G. Young, *J. Rech. Atmos.*, **15**, 17–37 (1981).
30. R. Mulvaney, G. F. Coulson and H. F. J. Corr, *Tellus*, **45B**. 179–187 (1993).
31. P. Artaxo, M. L. C. Rabello, W. Maenhaut and R. van Grieken, *Tellus*, **44B**, 318–334 (1992).
32. C. Boutron, *Atmos. Environ.*, **13**, 919–924 (1979).
33. W. H. Zoller, E. S. Gladney and R. A. Duce, *Science*, **183**, 198–200 (1974).
34. M. Murozumi, S. Nakamura and Y. Yoshida, *Mem. Natl. Inst. Polar Res.*, Special Issue, **7**, 255–263 (1978).
35. R. Chesselet, J. Morelli and P. Buat–Menard, *J. Geophys. Res.*, **77**, 5116–5131 (1972).
36. M. O. Andreae, R. J. Charlson, F. Bruynseels, H. Storms, R. van Grieken and W. Maenhaut, Science, **232**, 1620–1623 (1986)
37. M. R. Legrand, R. J. Delmas and R. J. Charlson, *Nature*, **334**, 418–420 (1988).
38. W. W. Berg and J. W. Winchester, in *Chemical Oceanography* (J. P. Riley and R. Chester eds.) Academic Press, London, vol. 7 (1978) pp 173–233.
39. R. Cini, N. Degli Innocenti, G. Loglio, G. Orlandi, A. M. Stortini, U. Tesei and R. Udisti, **63**, 15–27 (1996).
40. M. Legrand and R. J. Delmas, *Ann. Glaciology*, **7**, 20–25 (1985).
41. T. S. Bates, *J. Atmos. Chem.*, **14**, 315–337 (1992)
42. E. S. Saltzman, D. L. Savoie, J. M. Prospero and R. G. Zika, *J. Atmos. Chem.*, **4**, 227–240 (1986).
43. D. L. Savoie and J. M. Prospero, *J. Geophys. Res.*, **99D**, 3587–3596 (1994).
44. G. P. Ayers, J. P. Jvey and R. W. Gillet, *Nature*, **349**, 404–406 (1991).
45. G. P. Ayers, J. P. Ivey and H. S. Goodman, *J. Atmos. Chem.*, **4**, 173–185 (1986).
46. R. W. Gillet, G. P. Ayers, J. P. Ivey and J. L. Gras, *Dimethylsulphide: oceans, atmosphere and climate*, Restelli and Angeletti eds., Kluwer Acad. Publ., Dordrecht, 1993 p. 117–128.
47. B. C. Nguyen, N. Mihalopulos, J. P. Putaud, A. Gaudry, L. Gallet, W. C. Keene and J. N. Galloway, *J. Atmos. Chem.*, **15**, 39–53 (1992).
48. J. M. Prospero, D. L. Savoie, E. S. Saltzman and R. Larsen, *Nature*, **350**, 221–223 (1991).
49. A. J. Hynes, P. H. Wine and D. H. Semmes, *J. Phys. Chem.*, **90**, 4148–4156 (1986).

50. H. Berresheim, *J. Geophys. Res.*, **92D**, 13245–13262 (1987).
51. B. C. Nguyen, C. Bergeret and G. Lambert, *Gas Transfer at Water Surface*, W. Brunsaert and G. H. Jirka, 1984, D. Riedel,Hingham Mass.
52. E. S. Saltzman, NATO Adv. Res. Workshop, Annecy (1993).
53. M. E. Hawley, J. N. Galloway and W. C. Keene, *Water, Air and Soil Pollution*, **42**, 87–102 (1988).
54. R. Mulvaney, E. C. Pasteur, D. A. Peel, E. S. Saltzman and P.-Y. Whung, *Tellus*, **44B**, 295–303 (1992).
55. G. Piccardi, R. Udisti and F. Casella, *Dimethylsulphide: oceans, atmosphere and climate*, Restelli and Angeletti eds., Kluwer Acad. Publ., Dordrecht, 1993 p. 219–234.
56. D. L. Savoie, J. M. Prospero, R. J. Larsen and E. S. Saltzman, *J. Atmos. Chem.*, **14**, 181–204 (1992).
57. M. Legrand and C. Saigne, *Atmos. Environ.*, **22**, 1011–1017 (1988).
58. M. Legrand, C. Feniet–Saigne, E. S. Saltzman and C. Germain, *J. Atmos. Chem.*, **14**, 245–260 (1992).
59. S.-M. Li, L. A. Barrie, R. W. Talbot, R. C. Harris, C. I. Davidson and J.-L. Jaffrezo, *Atm. Environ.*, **27A**, 3011–3024 (1993).
60. A. J. Charlson, J. E. Lovelock, M. O. Anreae and S. G. Warren, *Nature*, **326**, 655–661 (1987).

AIR-SEA EXCHANGE: SEA SALT AND ORGANIC MICROCOMPONENTS IN ANTARCTIC SNOW

R. CINI, N. DEGLI INNOCENTI, G. LOGLIO, C. OPPO*, G. ORLANDI, A. M. STORTINI, U. TESEI and R. UDISTI*

*Laboratorio di Chimica Fisica Tecnica, Dipartimento di Chimica Organica "Ugo Schiff", Via Gino Capponi n.9, 50121, Firenze, Italy; *Dipartimento di Sanità Pubblica, Epidemiologia e Chimica Analitica Ambientale, Sezione Chimica Analitica, Via Gino Capponi n.9, 50121, Firenze, Italy*

A characterization of surface active fluorescent organic matter (SAFOM) in Antarctic snow is carried out. Its Fulvic Acids (FA) nature is confirmed. Its enrichment in the smallest aerosol particles is shown.

A tentative explanation of the presence of both natural and man-made organic microcomponents (SAFOM-interacting) is given in terms of marine aerosol transport. Their enrichment ratio appears of the same order as that of SAFOM, and their presence in the atmospheric particulate of marine origin supports the hypothesis on the transport of microcomponents in Antarctica "via marine aerosol".

KEY WORDS: Antarctic snow, air-sea interaction, marine aerosol, humic components, biogeochemical fractionation, pollutants.

INTRODUCTION

The long range transport of pollutants in remote areas via an "atmospheric path" is of increasing interest and so the object of many recent studies. The first results from the SEAREX project[1] evidenced the importance of the air-sea interaction in long range transport phenomena and showed how, for large areas, in the northern hemisphere the vapour phase appears to be the main state for most organics. These studies covered a vast area, including the Arctic and equatorial regions (Pacific Ocean)[1]. Evidence of long range transport of organic pollutants in Arctic regions, in the vapour state has also been shown recently[2–4].

On the other hand, marine aerosol is now seen as one of the most important sources of atmospheric particles on a global scale[1,5] that highly contributes to the nucleating phenomena of clouds and atmospheric precipitations[6].

Theoretical[7,8], laboratory[9,10] and field studies[11–13] have shown that the surface active matter and all those components able to interact with surfactants (both of natural and synthetic origin), are involved in the marine microlayer, in aerosol formation and in transport processes. The enrichment of such matter can reach many orders of magnitude with respect to its concentration in marine bulk water[12,13]

Antarctica, owing to its geographical situation, constitutes a site where the oceanic impact on the continent can be realized. Therefore, studies regarding the formation and transport phenomena of marine aerosol appear particularly attractive.

The evaluation of the organic matter as responsible for the association of other microcomponents in marine aerosol requires, in a first instance, an accurate analysis of its nature. The properties and the ability of the organic matter to distribute in particles of marine aerosol of different sizes must be studied in such conditions that the influence of other factors (i.e. mixing with aerosols of other nature and atmospheric chemistry effects) is reduced. These conditions are reached for rough sea with high coverage of whitecaps. Therefore, Antarctic salt storms[14], together with marine snow precipitations, represent the most favourable conditions. As a consequence, coastal Antarctic snow collected at different heights above sea-level represents interesting sample.

In addition, the study of the transport of pollutants toward Antarctica appears very important to ascertain the increase of certain contaminants in the southern hemisphere. According to us, the organic matter present in marine aerosol could be an important vehicle. This last aspect, that up to now has been largely underestimated, will be particularly addressed in this paper.

In previous studies[5,15–18], we have shown that surface active fluorescent organic matter (SAFOM) associated to marine aerosol is present in Antarctic snow. Further confirmation of the large presence of SAFOM in snow (for a major part of fulvic acid nature) could give an explanation for the anomalous excess of minor components in the Antarctic marine aerosol. The organic microcomponents interaction with SAFOM is considered here.

First of all, we characterize the SAFOM present in the snow by analysing a large sample and we discuss it in connection with the characterization of the samples collected in previous campaigns. Second, we try to evaluate the enrichment of natural and anthropogenic organics in the aerosol, using the sea salt as internal reference, in order to compare organics enrichment in the aerosol with the enrichment found for the SAFOM. Our aim is to verify if the hypothesis of transport of these organics by marine aerosol is correct.

EXPERIMENTAL

Sampling sites of snow and marine water

In the 1991/92 Italian Antarctic Campaign the sampling site for the surface snow was Mt. Melbourne (Lat. 74° 26' S, Long. 164° 45' E) at 1130 m above sea-level. In the same campaign seawater was sampled in a coastal station named "Faraglioni" near the Italian base in Terra Nova Bay.

In the 1990/91 Italian Antarctic Campaign, coastal water samples were taken during a short cruise of the "Cariboo" oceanographic ship in Terra Nova Bay (January, 13 1991). The coordinates of the sampling sites are given elsewhere[16].

Sampling and conservation technique

Snow. The surface snow (30 dm^3) was sampled by means of a Teflon blade. The present study included the analysis of a very large sample (60 dm^3) of surface snow

(5–8 cm depth) which should be referred to one of the last (spring/summer-beginning 1991) snow precipitations. The large sampling area could bring to an integration of the possible deposition differences of the surface marine salt that could occur in presence of a typical "salt storm" condition. The large sample permitted also further improvements in the treatment techniques employed for its final characterization.

Seawater. After the sampling, the water was stored in clean polyethylene containers and fast frozen to –70°C. The storage temperature was –30°C. The sampling and conservation techniques of the samples were described in detail in previous papers[15,16].

Procedures performed on the samples

On seawater and melted snow samples, we performed the non-foaming gas-bubble enrichment process as previously described[15]. This process mimics the first step of the natural aerosolization process taking place in the sea. As a result, we obtain two fractions: the enriched and the depleted fraction. The depleted fraction is given by the water remaining in the gas-bubble enrichment column after the process. This water is consequently "depleted" in its surface active components. The two fractions (enriched and depleted) and the untreated water, are then filtered in a Sartorius apparatus on a 0.45 μm Nuclepore membrane.

Measurement techniques

The measurement techniques and apparatus have been described elsewhere[2,5,15–18].

Fluorescence spectra. Emission and synchronous spectra were scanned with a Perkin Elmer Fluorescence Spectrophotometer (mod. LS-50B). The intensity values for the emission spectra were normalised using the formula[19]:

$$I_n = I_{max}/I_{Raman} \times 100$$

Where I_n denotes the normalised intensity, I_{max} the maximum intensity, I_{Raman} the Raman scattering peak intensity.

FT-IR. FT-IR spectra were recorded on freeze-dried sample as previously described[18]. The IR apparatus was a Bio-Rad FTP 40-PC.

Dynamic surface tension. The surface dilatational and thermodynamic properties were evaluated by surface rheology measurements performed with the Time Resolved Surface Viscoelastometer (TRSV) apparatus as previously described[20,21].

Weighing procedures. After the filtering operations, performed with a Nuclepore 0.45 μm filter, the filter membranes were dried and then weighed under vacuum with a Cahn 2000 ultramicrobalance with the procedures described elsewhere[15].

Turbidity. Turbidimetric measurements were performed with a Hach Double Beam Turbidimeter (mod. 2100).

Ion chromatography. The determination of Cl and Na concentration was performed by ion chromatography with the techniques described by Piccardi *et al.*[22]

Other parameters

From the dry mass data, the following parameters were calculated:

– *suspended matter concentration* (PM);
– *mass balance percent excess* (MB%) of the particulated matter, in non-foaming gas-bubble enrichment process, which can be expressed as follows[16]:

$$MB\% = \{[(PDM_e + PDM_d) - PDM_{unt}]/PDM_{unt}\} \times 100$$

where PDM_e denotes the dry mass (under vacuum at 20°C) of the particulated matter obtained by filtration of the enriched fraction of the sample, PDM_d the dry mass of the particulated matter in the depleted fraction of the sample and PDM_{unt} the dry mass of the particulated matter in the untreated sample (all masses are referred to the whole fractions of the sample).

The *surface rheological properties* were expressed by:

the limit surface elastic modulus $\varepsilon_0 = d\gamma/d\log\Gamma$,

where γ = surface tension, Γ = surface concentration excess;
and the characteristic frequency $\omega_0 = (dc/d\Gamma)^2 D/2$,
where c = bulk concentration and D = diffusion coefficient.

From the fluorescence and Cl concentration data, the *enrichment ratio* (F_r) with respect to the seawater (sw) composition was calculated. As pointed out elsewhere[5], the enrichment ratio is expressed as: $F_r = (I_{n\,snow}/Cl_{snow})/(In_{sw.}/Cl_{sw.})$.

RESULTS AND DISCUSSION

Characterization of the surface active fluorescent organic matter (SAFOM) present in Antarctic snow

Fluorescence spectra. Marine humic substances show (in the range 320–580 nm) an emission spectra characterised by a large fluorescence band (350–450 nm), when they are excited with near UV light (308 nm)[23]. As the emission spectra show frequently very broad bands, in order to get further information, synchronous scan spectra[24,25] are widely used. In Figure 1 are reported respectively the synchronous scan fluorescence spectra obtained with a $\Delta\lambda$ of 20 nm, for the marine Antarctic water, and for the enriched fraction of the melted snow. These spectra show similar features to those previously reported[5,26].

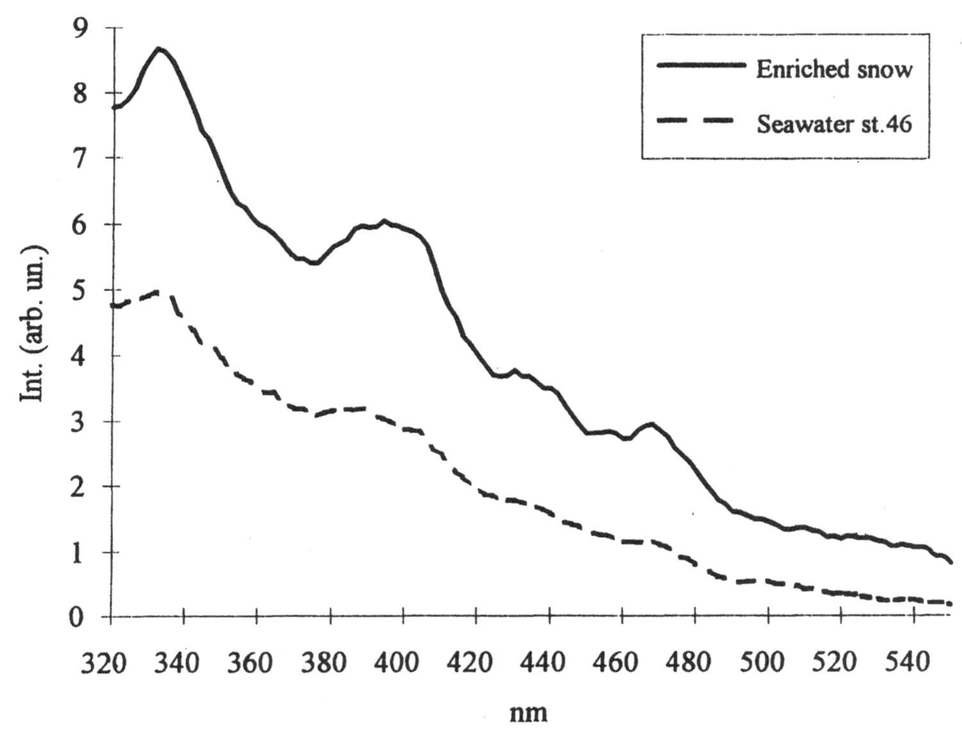

Figure 1 Synchronous scan spectra of the enriched fraction of the snow sampled at Mt. Melbourne station, and of surface Antarctic seawater sampled in station 46 (campaign 1990–91).

FT-IR spectra. In Figure 2 are reported respectively (a) the infrared spectra of marine fulvic acids (FA) extracted from the above reported sea surface water, and (b) the infrared spectra of the enriched melted snow after lyophilization. Because the lyophilization was done on the gas-bubble enriched fraction[15,16], a substantially larger mass of organic matter with respect to the saline component was obtained in the samples. In this case, the interference of SO_4^{2-} is lower than that in the previous spectra[17].

The comparison is done between a class of extracted compounds (FA) and the enriched fraction of all those organic substances, present in the snow, which include fatty acids and lipids in general (collected and selected principally because of their surfactant properties and molecular weight). Notwithstanding these limitations, the similarity appears well pronounced also for IR spectra. This strongly suggests the fulvic acid nature for a large part of the organics in the present sample, selected by the non-foaming gas-bubble extraction process[27]. The extraction and the separation of FA on a large sample (about 1 m^3) of Antarctic snow, collected during the 1993–94 campaign[26], confirm the above reported findings.

From the inspection of the IR spectra reported in Figure 2, independently from the interference of the SO_4^{2-} group, we cannot exclude the presence, in the enriched melted snow, of other surface active components which can have important additional role in the interaction and transport via "marine aerosol" of some particular components.

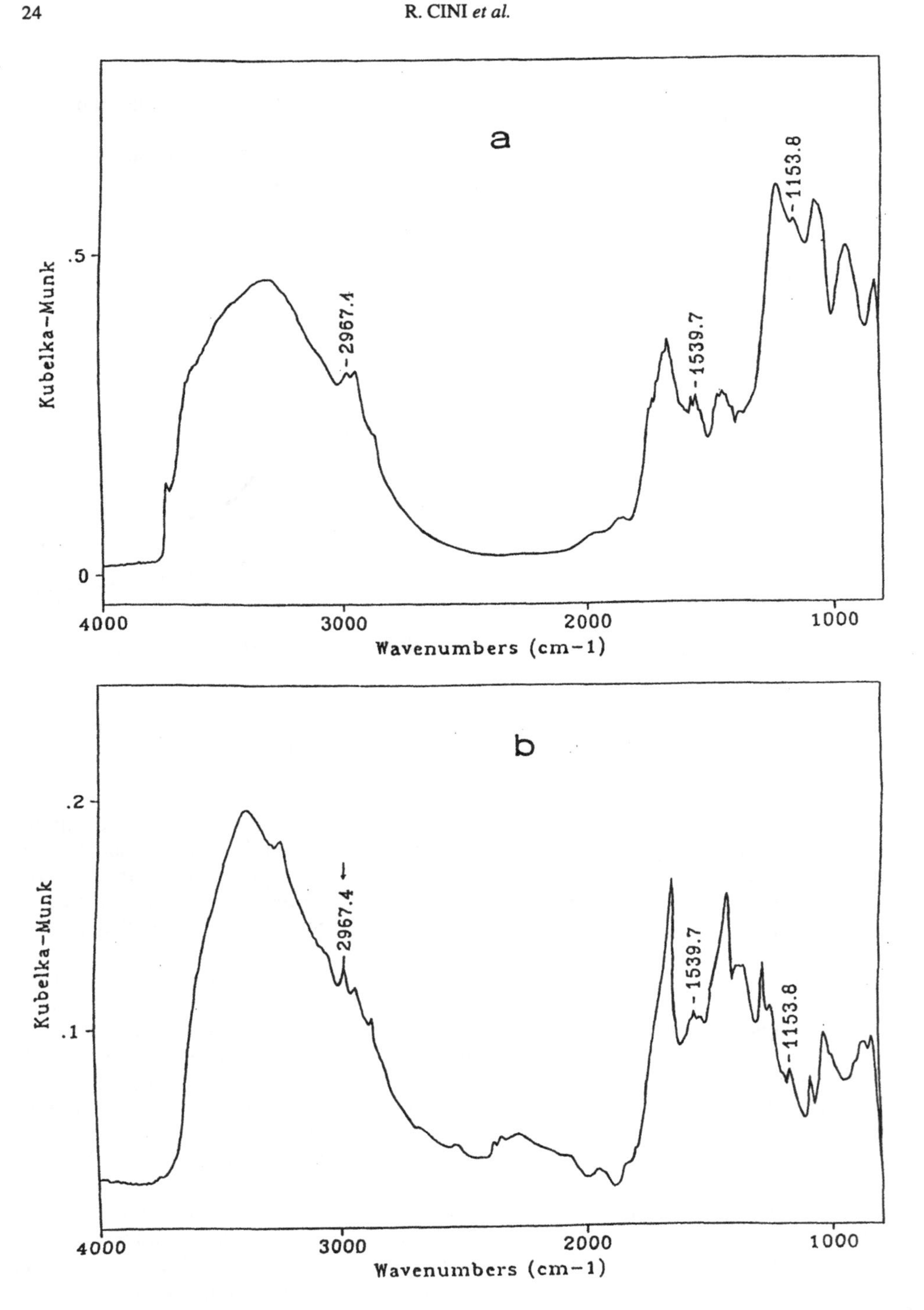

Figure 2 IR spectra of marine FA extracted from Antarctic surface seawater (campaign 1990–91) (a); enriched and lyophilised melted snow (same campaign) (b).

Characteristics of SAFOM and physico-chemical properties of natural samples

As reported by Chester[28], the microlayer at the air-seawater interface can be considered as a "soup" of many components in the largest part constituted of humic substances, polysaccharides, fatty acids, and their esters.

In the following, we will provide some data to facilitate the comparison of the SAFOM contained in seawater and in snow sampled in different campaigns. In Table 1 are reported the physico-chemical parameters of the large snow sample (1991–92 Antarctic campaign) together with those of the samples which were considered in other previous Antarctic campaigns[5,15,16].

In Table 2 are reported the mean values of the same parameters for sea surface water collected in 1990/91 both in a coastal station (6 samples), and during the "Cariboo" cruise in Terra Nova Bay (6 samples).

Surface active nature of organic matter in natural samples. MacIntyre[29] and Bock and Frew[30] suggest that the surface dilatational properties appear the most significant parameter for the physico-chemical characterization of surface active matter in seawater. As noted previously[15], surface dilatational properties could be able to characterise wet and dry surfactants. Taking into account the complex composition of SAFOM, Table 1 evidences, in particular for sample 4 (1991–92), the presence of surfactants as it results from ε_0 and ω_0 values. Moreover, for sample 4 the fitting of the experimental values of

Table 1 Physico-chemical properties of Antarctic melted snow (1988–1989, 1990–1991 and 1991–1992 campaigns)

Parameter	*Camp. 1988–89 snow 1130 m above sea-level* ①	*Camp. 1990–91 snow 1130 m above sea-level* ②	*Camp. 1990–91 snow 200 m above sea-level* ③	*Camp. 1991–92 snow 1130 m above sea-level* ④
Turbidity-NTU ± 1%	0.05	0.05	1.90	0.24
P.M. conc. $mg.dm^{-3}$ ± 1%	0.073	0.14	1.85	0.37
MB% ± 10%	+91	+62	−42	−23
I_n ± 2% (filtered)	traces	6.1	3.7	11.1
ε_0 mNm^{-1} ± 8%	0.2	21.0	21.5	20.7
ω_0 Hz ± 3%	n.d.	49×10^{-4}	51×10^{-4}	35.5×10^{-3}
Cl conc. $mg.dm^{-3}$ ± 1%	1.06	2.5	14.2	9.5

Sample ①, ② and ④ = snow, Station Mt. Melbourne (Lat. 74° 26' S, Long 164° 45' E) at 1130 m above sea-level, Sample ③ = snow, Station Carezza Lake (Lat. 74° 43' S, Long 164° 01' E) about 200 m above sea-level.

Table 2 Mean values of physico-chemical properties referred to six samples of Antarctic surface seawater (1990–91 campaign).

Parameter	*Mean*	*St. dev.*	*N*
Turbidity-NTU ± 1%	0.38	0.40	12
P.M. conc. $mg.dm^{-3}$ ± 1%	1.05	0.67	17
I_n ± 2%	29.75	11.34	17
ε_0 mNm^{-1} ± 8%	72.67	21.83	12
ω_0 Hz ± 3%	4.81×10^{-2}	4.75×10^{-2}	12

dilatational surface modulus to the diffusional model indicates the wet character of soluble surfactants, in agreement with the prevalent humic nature of the SAFOM. This aspect is reinforced by the large ω_0 value ($\omega_0 = 0$ is the characteristic frequency for an insoluble film).

Particulated matter (PM) and mass balance (MB%). Owing to the involvement of colloidal organic matter in the marine aerosol formation[15,16], the MB% information constitutes an interesting characterization parameter, utilised here for the first time for Antarctic snow. Referring to sample 4 (Table 1) it can be seen that the values of particulate matter, turbidity and In_0 are the highest with respect to samples 1 and 2. The interpretation of these data must be given in terms of seasonality, intensity and duration of meteorological events. In this respect, the snow sample 4 is probably referred to a "salt storm" event of stronger intensity than those occurring in samples 1 and 2. A more intense "salt storm" with higher winds could have transported particles of larger size at 1130 m above sea-level. Therefore, the negative mass balance excess (MB%) obtained in the non-foaming gas-bubble enrichment process should indicate that the sample was subjected to a lower oxidating UV action[16] as a consequence of the larger particle dimensions. This hypothesis is in agreement with the previous interpretation[5] given for sample 3 on the fluorescence enrichment and its negative MB%[18]. We may also observe that this finding is in line with particulated matter selection predicted according to the proposed models of MacIntyre and Blanchard[7,8] for the particles of greater dimensions.

Parameters related with transport phenomena: the fluorescent surfactants enrichment

The enrichment ratio F_r of SAFOM for sample 4 (Table 1) was found to be 1.3×10^{-3}. The In_{SW} of reference was taken the mean value of the seawater samples in all our campaigns (1987–88, 1988–89, 1990–91) which resulted $In_{SW} = 10.9$.

A number of studies have been reported in the literature which show experimentally that the ratio OM/Na increase with the dimensional lowering of the marine aerosol particles. We would like to stress the fact that the SAFOM amount increases with the decrease of the particle diameter, as suggested by the literature for the OM/Na, of which SAFOM makes a major part.

Hoffman and Duce[31] showed that about 80% of the particulated organic carbon (POC) was present in particles smaller than 2 μm. Barker and Zetlin[32] found the highest OM concentration in the smallest sampled particles. We also recently have found similar behaviour for the surfactant fluorescent matter (SAFOM) in samples collected at our meteomarine station in Leghorn (Tyrrhenian sea coast). We report here the samples collected in Leghorn for two reasons. First, the amount of material we have collected for any sampling operation in Antarctica, was not enough to make a comparison between different samples collected in particular weather conditions for different known particles dimensions. Second, we can consider the dependence of the SAFOM amount on the particles diameter during the enrichment and the aerosolization processes, as well as during the long range transport phenomena.

In Figure 3a are reported typical results for a marine aerosol integrated sample, sampled in rough sea conditions, with inland winds (185°–270° direction and speed > 6 m/s). An Andersen high volume sampler MP 1000 equipped with a Sierra 235 five stage impactor used in the standard flux, piloted with an intelligent meteorological trigger[33] permitted to reach a final volume of 12700 dm^3 comprehensive of many stormy events. Above each histogram column the Na concentrations, are reported, for the corresponding particle dimension. In Figure 3b are reported the corresponding F_r vs.

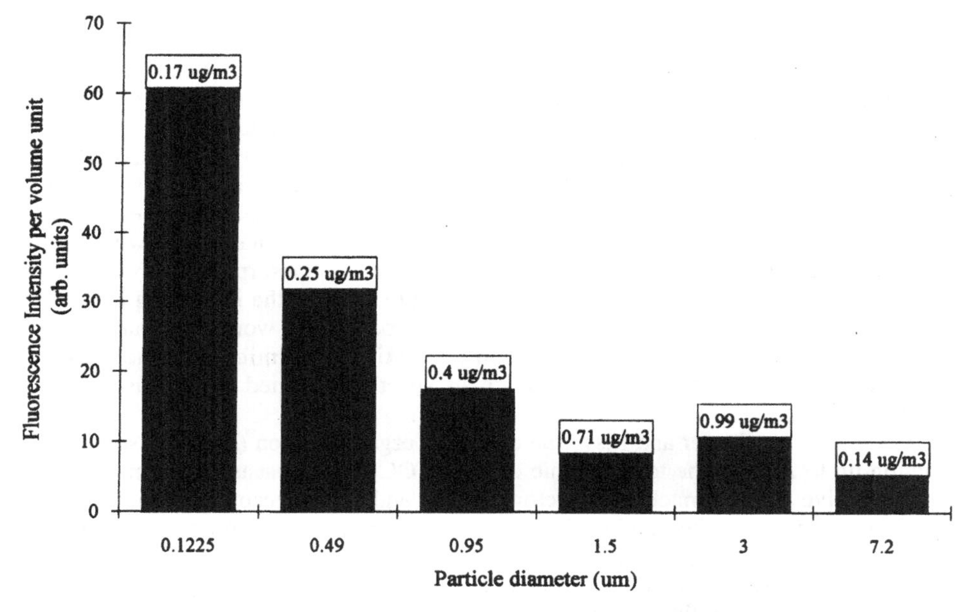

Figure 3a Histogram of the fluorescence intensity of SAFOM per unit volume vs. particle diameter for the Leghorn aerosol sample. Above each column are reported the Na concentrations of the aerosol.

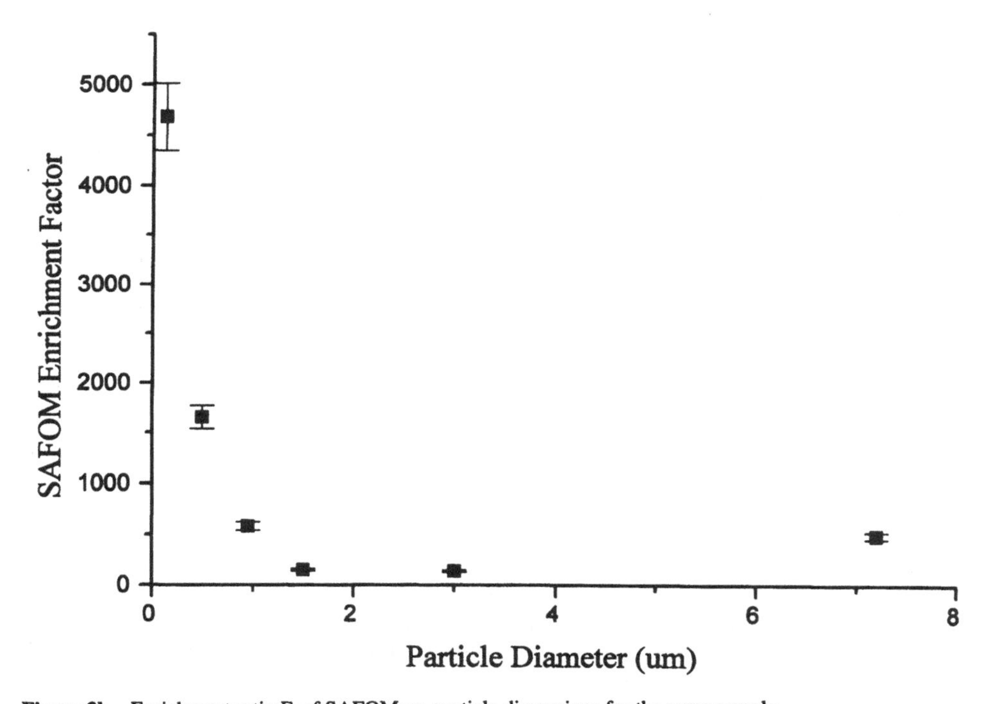

Figure 3b Enrichment ratio F_r of SAFOM vs. particle dimensions for the same sample.

particle diameter values, referred to the mean fluorescence intensity of the seawater samples collected in the same period.

These results, in first instance, evidence for the SAFOM a well defined increase, with the dimensional decrease of the aerosol particles. Therefore, they reinforce the interpretation on our previous findings on Antarctic snow SAFOM F_r[5] in terms of the dimensional selection of the particles due to the altitude effect. They also appear in agreement with our hypothesis that, for our last Antarctic snow sample, a strongest salt storm, should be responsible of particles injection with higher size particles. No direct comparison with Fr of Leghorn SAFOM with that present in Antarctic snow is possible, because the complex mixtures of components that form the adsorption microlayer of the two places are expected different. A further interpretation of the increasing SAFOM Fr with the dimensional lowering of particles will be reported in a work in preparation.

It is necessary here to recall the characteristics of humic substances and the characteristics of the sea microlayer[34] which are strictly correlated to some aspects of the aerosolization process[7,8].

As reported by Hunter and Liss[34] the dissolved organic carbon (DOC) in seawater is a fraction up to 20% of the total organic carbon (TOC) and consists for the most part, of surface active species much as planctonic lipids and proteinaceous materials. Stuemer and Harvey[35] isolated the humic substances from marine water and concluded that they constitute the most important contributor of film forming substances at the air-sea interface. Therefore, the air-sea interface -i.e. the marine microlayer- and the new interface originated by the breaking waves events, undergo an enrichment of the above mentioned "soup" of SAFOM[28], together with interacting microcomponents as natural and synthetic organic compounds, transition metals, and microorganisms[8], both in the dissolved and in the aggregate state[36]. The whole phenomena are well represented by the proposed scheme of Lion and Leckie[36].

Transport of organic microcomponents into the snow

In a previous work[10], it was evidenced the transport of many natural and synthetic organic components, with aerosolization experiments in laboratory, both on artificial and natural seawater samples. The presence in dissolved state of SAFOM was always detected together with other organics transferred from seawater to air. The presence of particulated matter depressed the transfer efficiency. Moreover, for some organics the partition between bulk water, aerosol spray, and surrounding atmosphere, subsequent to the non-foaming gas-bubble process, was evaluated in relation with their vapour pressure.

Desideri *et al.*[37–39] reported the concentration of many classes of natural and anthropogenic organic compounds in surface Antarctic water, ice pack and coastal snow.

The question arises if the organic products are present in Antarctic snow as a result of direct transport together with marine aerosol or of their vapour scavenging from the atmosphere by the snow precipitations.

Transport of organics in Antarctic snow. To check this point we evaluated the enrichment ratio F_r for four typical products. The selection of the organic compounds was done considering both natural and anthropogenic materials and their behaviour with respect to their surface physico-chemical properties and to their interactions with fulvic acids. Three stations were considered in two different campaigns, respectively: campaign 1987–88 for the Campbell Glacier (920 m above sea-level) and Mt. Crummer (900 m above sea-level) stations[38]; and campaign 1990–91 for Mt. Melbourne (1130 m above sea-level) station. The concentrations of the considered organic compounds in surface seawater were referred to the same campaign for the first two stations[38]. For the Mt. Melbourne station the enrichments on the snow samples were calculated, in a first approximation, referring to the concentration of the same compounds in seawater sampled in the 1988–89 campaign by the same authors[39]. The average values of Cl seawater concentration are taken from the literature[40].

In Tables 3a and 3b are reported the concentrations both in the melted snow and in the surface seawater for the above mentioned stations. In Table 3c are reported the enrichment ratios F_r as previously defined. Notwithstanding, the above mentioned limitations, the mean values for the snow sampled in the two campaigns appear consistent with those expected for organics interacting with marine aerosol, and with the SAFOM F_r found in Antarctic snow.

Table 3 Concentrations and enrichment ratios (F_r) of selected organics in Antarctic snow and seawater (in ng/dm^3) (from the data of Desideri *et al.*[37,38])

(a)

	Mt. Crummer 900 m above sea-level	*Campbell Glacier 920 m above sea-level*	*Seawater mean values 1987–88*
(1) C-31	b.d.l.	56	3.30
(2) Squalene	–	–	–
(3) Di-iso-butyl-phthalate	102	126	203
(4) Bis-2-ethyl-hexyl-phthalate	b.d.l.	88	129
Chlorine (Piccardi *et al.*)[22]	7.36×10^5	1.73×10^6	1.98×10^{10}

(b)

	Snow Mt. Melbourne 1990–91	*Seawater mean values 1988–89*
(1) C-31	37	9.30
(2) Squalene	121	22.7
(3) Di-iso-butyl-phthalate	61	22.1
(4) Bis-2-ethyl-hexyl-phthalate	173	51.2
Chlorine (our data)[16]	9.5×10^6	1.98×10^{10}

(c)

Enrichment ratio	*Mt. Melbourne 1990–91*	*Mt. Crummer 1987–88*	*Campbell Glacier 1987–88*
F_r (1)	8.2×10^3	4.03×10^4	2×10^5
F_r (2)	8.6×10^3	–	–
F_r (3)	6×10^3	1.3×10^6	7×10^3
F_r (4)	7×10^3	–	7.8×10^3

b.d.l.—Below detection limits

The presence of anthropogenic components in Antarctic aerosol of marine origin. The recent finding on the Antarctic atmospheric particulates collected by Ciccioli *et al.*[41] show that samples collected prevalently from marine air masses are particularly enriched in organics, including phthalates, that, for their physico-chemical properties, are well involved in the process of marine aerosol formation[10]. The authors evidence that many of synthetic organics found in their atmospheric particulate samples are the same as those found in marine Antarctic water, ice pack, and snow[37–39]. Taking into account the predicted interactions of these compound with humic substances, and their F_r values found in the snow, these results further support the hypothesis that during the "salt storms", many organic compounds present in surface seawater are associated to marine aerosol, both as a result of their surface properties and of their interaction with SAFOM. Therefore, a large part of those synthetic organic compounds found by Desideri *et al.*[39] in Antarctic snow should be prevalently associated with the marine aerosol transport, as also our recent laboratory experiments showed[10].

These results could be considered of particular relevance because they are referred to the Antarctic summer period in which the maximum concentration at the vapour state of organic compounds is to be expected.

CONCLUDING REMARKS

The presence in Antarctic snow of fluorescent surface active matter of marine origin is confirmed. The fulvic acids (FA) nature of this matter reinforces its role in the transport phenomena via "marine aerosol" in Antarctica.

The presence of some organic pollutants in Antarctic snow could therefore be explained in terms of transport via this "marine aerosol". The support to this last conclusion is given by the enrichment ratio values here reported, in agreement with those found by measurements of surface active fluorescent organic matter (SAFOM) of fulvic acid nature in Antarctic snow at similar altitude on the sea-level and laboratory experiments following the marine aerosol formation models of MacIntyre and Blanchard.

Acknowledgements

We are indebted with Prof. Piccardi and his co-workers for the fruitful discussions. We thank also ENEA for the logistic assistance. We thank also Dr. Bellandi for its ion chromatography detection on Leghorn aerosol sample.

This work was supported by PNRA, Chemical Methodologies for Environmental Impact Section.

References

1. R. A. Duce, in: *Chemical Oceanography* (J. P. Riley and R. Chester, eds., Academic Press, London N.Y., Vol. 10, 1989) pp. 1–14.
2. M. Oehme and S. Manø, *Fresenius Z. Anal. Chem.*, **319**, 141–146 (1984).
3. J. M. Pacyna and M. Oehme, *Atmos. Environ.*, **22**, 243–257 (1988).
4. M. Oehme, *Ambio*, **20**, 293–297 (1991).
5. R. Cini, N. Degli Innocenti, G. Loglio, G. Orlandi, A. M. Stortini and U. Tesei, *Intern. J. Environ. Anal. Chem.*, **55**, 285–296 (1994).
6. D. C. Blanchard and A. H. Woodcock, *Tellus*, **9**, 145–158 (1957).
7. F. MacIntyre, *J. Geophys. Res.*, **77**, 5211–5228 (1972).

8. D. Blanchard, *Adv. Chem. Ser.*, **145**, 360 (1975).
9. E. J. Hoffman and R. Duce, *J. Geophys. Res.*, **81**, 3667–3669 (1976).
10. R. Cini, P. G. Desideri and L. Lepri, *Anal. Chim. Acta*, **291**, 329–340 (1994).
11. E. J. Hoffman and R. Duce, *J. Geophys. Res.*, **79**, 4474–4477 (1974).
12. F. C. Cattel and W. D. Scott, *Science*, **20**, 429–430 (1978).
13. R. Cini, C. Oppo, S. Bellandi, G. Loglio, N. Degli Innocenti, A. M. Stortini, E. Schiavuta, U. Tesei, G. Orlandi and F. Pantani, In: *Atti del I° Simposio Nazionale sulle "Strategie e Tecniche di Monitoraggio sull'Atmosfera", Roma 20–22 settembre 1993* (P. Ciccioli ed., SCI, Rome, 1993) pp. 168–175.
14. G. E. Shaw, *Rev. Geophys.*, **26**, 86–112 (1988).
15. G. Loglio, N. Degli Innocenti, U. Tesei, A. M. Stortini and R. Cini, *Ann. Chim. (Rome)*, **79**, 571–587 (1989).
16. G. Loglio, N. Degli Innocenti, A. M. Stortini, G. Orlandi, U. Tesei, P. Mittner and R. Cini, *Ann. Chim. (Rome)*, **81**, 453–467 (1991).
17. R. Cini, N. Degli Innocenti, G. Loglio, P. Mittner, A. M. Stortini and U. Tesei, In: *IV Workshop of Italian Research on Antarctic Atmosphere* (M. Colacino, G. Giovanelli and L. Stefanutti eds., SIF, Bologna, 1992) Conference Proceedings Vol. 35 pp. 191–203.
18. R. Cini, N. Degli Innocenti, G. Loglio, A. M. Stortini and U. Tesei, In: *V Workshop of Italian Research on Antarctic Atmosphere* (M. Colacino, G. Giovanelli and L. Stefanutti eds., SIF, Bologna, 1994) Conference Proceedings Vol. 36 pp. 193–206.
19. G. Nyquist. "Investigation of some optical properties of seawater with special reference to lignin sulphonates and humic substances". *PhD. Thesis*, Marine and Analytical Chemistry Dept., Göteborg University (1979).
20. G. Loglio, U. Tesei and R. Cini, *Rev. Sci. Instr.*, **59**, 2045–2050 (1988).
21. G. Loglio, U. Tesei and R. Cini, *Boll. Oceanolog. Teor. Appl.*, **4**, 91–96 (1986).
22. G. Piccardi, R. Udisti and E. Barbolani, *Ann. Chim. (Rome)*, **79**, 701–702 (1989).
23. T. M. Miano, G. Sposito and J. P. Martin. *Soil Sci. Soc. Am. J.*, **52**, 1016–1019 (1988).
24. Tuan Vo-Dinh, *Anal. Chem.*, **50**, 396–401 (1978).
25. J. B. F. Lloyd, *Nature (London)*, **231**, 64 (1971).
26. R. Cini, B. M. Petronio, N. Degli Innocenti, A. M. Stortini, C. Braguglia and N. Calace, *Ann. Chim. (Roma)*, **84**, 424–429 (1994).
27. G. Loglio, U. Tesei, P. Cellini Legittimo, E. Racanelli and R. Cini, *Ann. Chim. (Rome)*, **71**, 251–262 (1981).
28. R. Chester, *Marine Geochemistry* (Unvin Hyman Ltd. Boston Sydney Wellington, 1990). p. 125–129.
29. F. MacIntyre, *J. Recherche Atmospherique*, **8**, 515 (1974).
30. E. J. Bock and N. M. Frew, *J. Geophys. Res.*, **98**(C8), 14,599–14,617 (1993).
31. E. J. Hoffman and R. Duce, *J. Geophys. Res. Lett.*, **4**, 449–452 (1977).
32. D. R. Barker and H. Zetlin, *J. Geophys. Res.*, **77**, 5076–5086 (1972).
33. E. Schiavuta, P. Mittner and R. Cini, *Workshop Il Monitoraggio Automatico dell'Inquinamento Marino" CNR-SCI—Taranto 9–10 Aprile 1992,* (N. Cardellicchio and F. Dell'Erba eds., SCI-CNR, 1992), pp. 313–330.
34. K. A. Hunter and P. S. Liss, In: *Marine Organic Chemistry* (E. K. Duursma, L. R. Dawson eds.), Elsevier Publ., Amsterdam, pp. 259–298 (1981).
35. D. Stuermer and G. Harvey, *Deep-Sea Res.*, **24**, 303–309 (1977).
36. L. W Lion and J. O. Leckie, *Environ, Geol.*, **3**, 293–314, (1981).
37. P. G. Desideri, L. Lepri, L. Checchini and D. Santianni, *Intern. J. Environ. An. Chem.*, **55**, 33–46 (1994).
38. P. G. Desideri, L. Lepri and L. Checchini, *Ann. Chim. (Rome)*, **79**, 589–605 (1989).
39. P. G. Desideri, L. Lepri and L. Checchini, *Ann. Chim. (Rome)*, **81**, 295–416 (1991).
40. *Chemical Oceanography* (J. P. Riley and G. Skirrow, eds., Academic Press, London N. Y., Vol. 1, 1975) p. 558.
41. P. Ciccioli, A. Cecinato, E. Brancaleoni, M. Montagnoli and I. Allegrini, *Intern. J. Environ. Anal. Chem.*, **55**, 47–59 (1994).

DISTRIBUTION OF ^{226}Ra IN THE ROSS SEA—ANTARCTICA

M. G. BETTOLI[1], L. CANTELLI[1], G. QUEIRAZZA[2], M. ROVERI[2], L. TOSITTI[1], O. TUBERTINI[1] and S. VALCHER[1]

[1]*Environmental Radioactivity Center, Dept. of Chemistry, University of Bologna v. Selmi 2, 40126 Bologna, Italy;* [2]*ENEL S.p.A.—Center for the Environmental and Material Research, v. Monfalcone 15, 20132 Milan, Italy*

An improved procedure for the determination of ^{222}Rn and ^{226}Ra in seawater developed for easier on-board operations is presented. Data on the radioactive disequilibrium between the mentioned radionuclides as directly determined in the field can be greatly helpful in the study of the gas-exchange processes at the air-sea interface, especially as far as the Antarctic Ocean is concerned. The method employed has been preliminarly tested in laboratory on a set of seawater samples collected in the Ross Sea and ^{226}Ra data collected are discussed and compared to literature data.

KEY WORDS: ^{226}Ra, ^{222}Rn, Antarctic ocean, liquid scintillation counting.

INTRODUCTION

Gas exchange at air-sea interface is a process that plays a key role within the dynamics of global biogeochemical cycles, in particular as regards the evaluation of problems of relevant environmental concern such as the "greenhouse effect", related to the increasing average concentration of reactive gases such as CO_2, CH_4 etc. in the troposphere[1,2].

In order to estimate fluxes between the atmosphere and the ocean, it is necessary to determine the air-sea transfer coefficient. One of the methods employed is based upon the determination of ^{222}Rn deficit in respect to ^{226}Ra along the water column. The inert character of the ^{222}Rn ($t_{1/2}$ = 3.8 d) makes it suitable for a study of this important problem. ^{226}Ra in the deep oceans is generally found to be in radioactive equilibrium with the daughter ^{222}Rn, but in near-surface seawater this can escape by diffusion into the atmosphere, due to its lower activity in air than in seawater, causing a disequilibrium with respect to ^{226}Ra[3–6].

In addition, ^{226}Ra itself can be used effectively as a tracer of seawater masses, as was pointed out by Koczy[7]. The presence of ^{226}Ra ($t_{1/2}$ = 1600 y) in seawater is due to its primary deep-sea source by ^{230}Th decay in sediments; high inventories of thorium are maintained in the upper sediment column because this isotope is rapidly removed from seawater after its production by ^{234}U decay.

For the large variation of ^{226}Ra on a global scale, this natural radionuclide has been employed, as one of the most important radioactive tracers, in the study of large-scale mixing and ocean circulation, which have been the main effort of the GEOSECS (Geochemical Ocean Sections Study) program between 1970 and 1980[8–14].

In the framework of the Italian Program of Research in Antarctica and taken into account SCAR suggestions for primary scientific requirements[15], an investigation has been recently undertaken with the aim of evaluating the absorption capacity of the Antarctic waters towards the ever increasing concentration of reactive gases as CO_2, CH_4, etc. in the troposphere.

In the present work we describe a new experimental procedure for the determination of ^{226}Ra through ^{222}Rn in seawater. The method described has been set up and tested by processing seawater samples collected in the Ross Sea during the 1988–89, 1989–90 and 1990–91 Italian expeditions in Antarctica.

In spite of the limited number of samples available, some considerations on ^{226}Ra distribution are presented and are compared with those described in the literature.

EXPERIMENTAL

Properties of radon and radium

As a noble gas, radon shares with its homologues reduced chemical reactivity, a remarkable gaseous diffusivity. It is considered as the major source of radioactive disequilibrium in the uranium and thorium decay series and it has an appreciable water solubility (about 0.5 g/l at STP). All radon isotopes are radioactive; the longest-lived ^{222}Rn ($t_{1/2}$ = 3.8 d) tends to behave conservatively in the water column, because its concentration in seawater is supported by that of its radium parent with which it is usually in radioactive equilibrium.

Radium is an alkaline earth element with chemical properties similar to barium, it is an electropositive element and tends to form strong ionic bonds. Most Ra salts are insoluble, particularly sulphates and carbonates. Among the various isotopes of the element, all radioactive, ^{226}Ra, ^{230}Th daughter, has the longest half-life (1600 y).

Analytical procedure

Radium can be determined either directly through its coprecipitation with barium and following α or γ-spectrometry, or mostly by the well-known "radon emanation" technique, in which ^{226}Ra in solution is measured indirectly through its daughter ^{222}Rn once stripped the radon initially present and subsequently reached radioactive equilibrium in a sealed container.

The method employed in this work has been developed in order to be able to determine both radionuclides through ^{222}Rn[16,17]. The improvement of the methods usually described in the literature[4,16], for the quantitative determination of ^{226}Ra in the field, is based on the introduction of two new steps: modified ^{222}Rn desorption[18] and ^{226}Ra preconcentration.

The main steps of the complete experimental procedure are described below.

Sampling

Surface seawater was sampled by a membrane and submersible pumps, while 30 l Niskins bottles were employed for vertical profiles. The samples were stored in polyethylene tanks kept frozen at –30°C. Geographical location of sampling stations are shown in Figure 1.

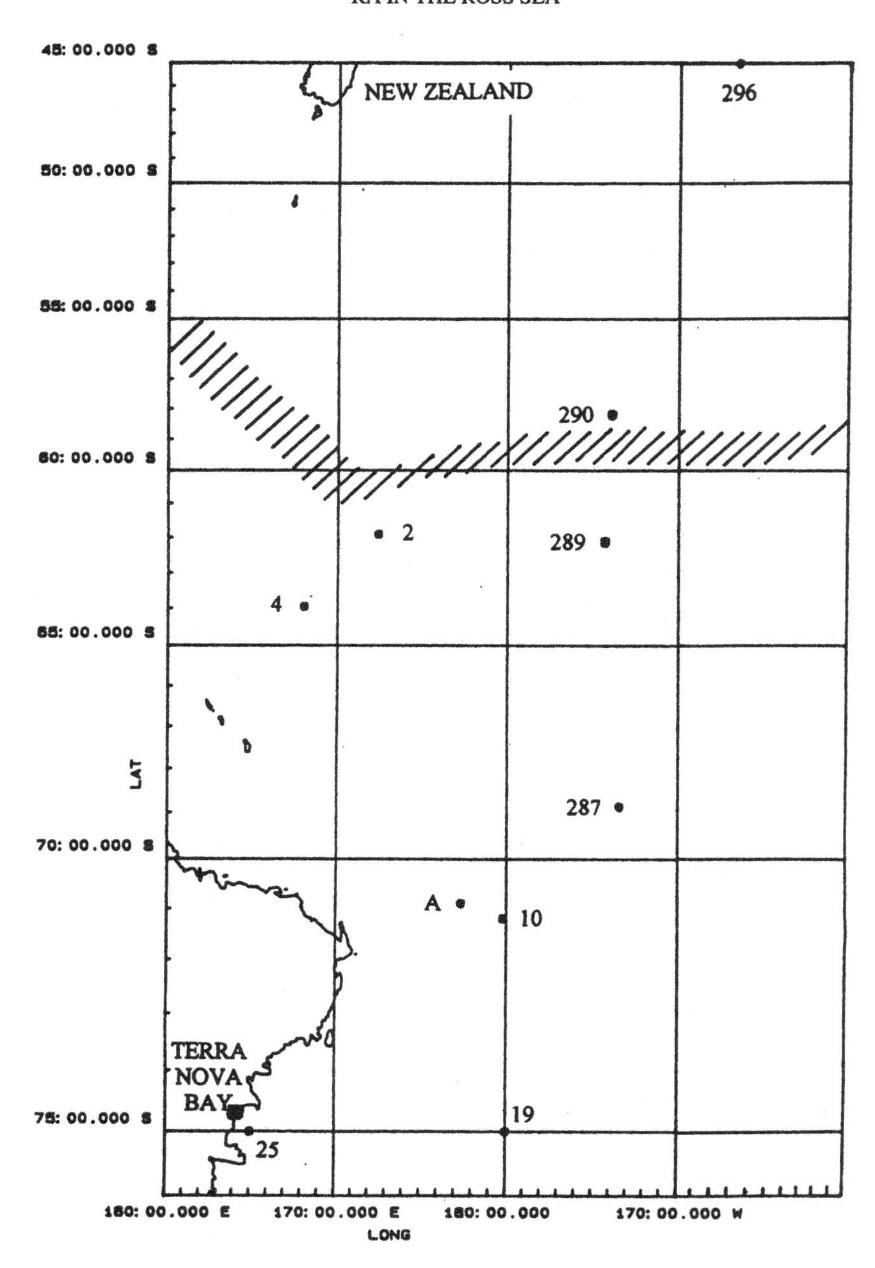

Figure 1 Locations of the ^{226}Ra station reported in this study and those reported previously by Ku *et al.*[9] and by Peng *et al.*[4]. The approximate position of the Antarctic Convergence is indicated (N.B.: station 2SW, is located at 5 km from the coast, near the Italian Base at Terra Nova Ray).

Separation and preconcentration of ^{222}Rn

Water samples were melted and transferred into airtight stainless steel (AISI 316L) 28 l tanks, internally lined with epoxy resin and left for at least one month to allow for the radioactive equilibrium between ^{226}Ra and ^{222}Rn[17].

Radon was then stripped with helium and recirculated for 1 hour in an airtight handling system by means of a membrane pump (flowrate: 2 l/min). Humidity and CO_2 were trapped by ascarite and silica gel and the extracted radon was adsorbed onto a column (Φ = 8 mm; 1 = 260 mm) filled with 5 g of active coconut charcoal with a low ^{226}Ra activity (< 1 Bq/kg, 10–35 mesh, Fisher Chemical). The column was cooled down to –86°C with an acetone/dried ice mixture. The possible loss of radon is negligible since its boiling and melting points are –65°C and –71°C respectively.

^{222}Rn desorption

The refrigerated column was completely filled with toluene in countercurrent (0.3 ml/min) in respect to previous adsorption flux. It was removed from the refrigerating bath and left reaching room temperature; elution was carried out with 12 ml of toluene in a scintillation counting glass vial after addition of 4 ml of a suitable scintillation fluor (PPO: 4 g/l and POPOP: 0.5 g/l in toluene). Recovery of radon not retained in the first vial was achieved by means of a second vial containing extra 4 ml of toluene. Finally the contents of both vials were combined for subsequent Liquid Scintillation counting.

Liquid Scintillation counting conditions

Scintillation counting was effected at least three hours after sample preparation to allow for further equilibration between radon and its own daughters according to the decay chain, showing that each radon decay is followed by the emission of five particles:

	α		α		β,γ		β,γ		α
^{222}Rn	$\rightarrow$	^{218}Po	$\rightarrow$	^{214}Pb	$\rightarrow$	^{214}Bi	$\rightarrow$	^{214}Po	$\rightarrow$
	3.82	Radium A	3.04 m	Radium B	26.8 min	Radium C	19.9 min	Radium C'	164 µs

Counting was carried out with a total α–β counter model TriCarb 1000 (Canberra Packard) allowing for spectra acquisition on a portable P.C. by Canberra Packard Datalink and Spectrograph software packages. In order to improve sensitivity in radon measurement, the counting window was optimized so that to obtain a low background and high efficiency; this has allowed to optimize the instrumental figure of merit expressed as the ratio between the square of the counting efficiency and the background.

For given chemical extraction and elution yields (respectively Y_d and Y_e), counting efficiency (E in counts s^{-1} Bq^{-1}), sample volume (V in m^3), sample and background count rate (respectively c_s and c_b in count s^{-1}), constant of decay of radon ($\lambda = 2.1 \cdot 10^{-6}$ s^{-1}), counting sample time (t = 7200 s), time elapsed after radon stripping and time of measurement (Δt), the activity of radon (A in Bq/m^3) was determined according to the following expression:

$$A = \frac{c_s - c_b}{5 \cdot V \cdot Y_d \cdot Y_e \cdot E} \cdot \frac{\lambda \cdot t}{(1 - e^{-\lambda t})} \cdot \frac{1}{(e^{-\lambda \Delta t})}$$

The standard deviation of A was evaluated as:

$$s = \frac{\{[A \cdot 5 \cdot V \cdot Y_d \cdot Y_e \cdot E \cdot e^{-\lambda \Delta t} (1 - e^{-\lambda t})/\lambda t \cdot t] + 2\, c_b \cdot t\}^{0.5}}{5\, V \cdot Y_d \cdot Y_e \cdot E \cdot e^{-\lambda \Delta t} (1 - e^{-\lambda t})/\lambda t \cdot t}$$

which is assumed to be the best estimate for the population standard deviation (σ). The values of A are reported at a confidence limit of 95% (2σ).

Under the experimental conditions adopted, the Lowest Detectable Activity (LDA expressed in Bq/m^3) is evaluated by the expression[19]:

$$LDA = \frac{1 + (1 + 2\, c_b \cdot t)^{0.5}}{0.5\, V \cdot E \cdot t \cdot 5 \cdot Y_d \cdot Y_e}$$

The suitability of the experimental procedure described was tested both by laboratory and field work. In particular, lab-tests were carried out by means of seawater samples depleted in natural ^{226}Ra by passing through a MnO_2-impregnated resin column, subsequently spiked with standard solutions of ^{226}Ra (NBS, 4953D, 147.5 Bq/kg) in order to accurately evaluate the overall recovery efficiency of the method, together with the radiometric efficiency. The total efficiency of this method is 65%.

The values of LDA for the two radionuclides at different volumes of the sample and under different working conditions are reported in Table 1.

^{226}Ra preconcentration method

100 1 samples were eluted through a column (ϕ = 25 mm; 1 = 300 mm) filled with 100 ml of Amberlite XAD – 7 (20–50 mesh, BDH Chemicals LTD.) MnO_2 impregnated. After adsorption, the resin was treated with hot 8 M HCl to total dissolution of MnO_2. The hydrochloric solution was then allowed to stay for radioactive growth of ^{222}Rn till equilibrium with ^{226}Ra was reached (1 month), using a 250 ml glass bubbler equipped with airtight stopcocks. Radon was finally extracted as previously described.

Table 1 LDA (Bq/m^3) for ^{226}Ra and ^{222}Rn in seawater.

	Volume sample (m^3)	*LDA on board*	*LDA laboratory*
^{226}Ra	0.10	0.05	0.07
^{222}Rn	0.028	0.19	0 25

RESULTS AND DISCUSSION

The validity of the method has been confirmed during oceanographic cruises carried out in the Mediterranean Sea by ENEL (Milano)[16].

The availability of seawater samples from past Antarctic expeditions has offered the opportunity to apply this method for obtaining preliminary results useful for the investigations in the future oceanographic expeditions in Antarctica in sampling stations of the Ross Sea.

Data presented refer to ^{226}Ra activity in Bq/m^3 (1 Bq/m^3 of ^{226}Ra is equivalent to $2.8 \cdot 10^{-13}$ g/l) in surface seawater samples, collected at a depth ranging between 1 and 3 m (Table 2) and in the water column (Table 3). As a matter of comparison GEOSECS data

Table 2 Activities of ^{226}Ra (Bq/m^3) in surface seawater samples in the Ross Sea and comparison with literature data from the same area[4,9].

Station N°	*Lat.*	*Long.*	*Sample size (l)*	*Bottom depth (m)*	*^{226}Ra*
296 Geosecs (1974)	*44°59'S*	*166°42'W*		*5340*	*1.25 ± 0.10*
290 Geosecs (1974)	*58°00'S*	*174°00'W*		*5177*	*2.55 ± 0.10*
2 (1989–90)	61°49'S	172°38'E	100	3850	2.22 ± 0.10
289 Geosecs (1974)	*62°00'S*	*174°00'W*			*2.38 ± 0.10*
4 (1989–90)	63°58'S	168°02'E	100	2980	2.06 ± 0.11
287 Geosecs (1974)	*69°05'S*	*173°30'W*		*4149*	*2.83 ± 0.10*
A (1990–91)	70°53'S	177°21'E	25	1800	2.37 ± 0.25
10 (1989–90)	71°12'S	179°50'E	100	1500	2.40 ± 0.11
19 (1989–90)	75°01'S	179°59'E	25	460	2.99 ± 0.25
25 (1989–90)	75°01'S	170°00'E	100	940	2.98 ± 0.18
2 SW (1988–89)(a)	74°41'S	164°18'E	25	300	3.64 ± 0.29
7 SW (1988–89)(a)	74°41'S	164°18'E	25	300	3.35 ± 0.28

(a) Same station before (2SW) and after (7SW) pack melting.

Table 3 Activities profile of ^{226}Ra (Bq/m^3) and comparison with literature data from the same area[9].

Station 10 (1989–90) (71°12'S, 179°50'E) bottom depth: 1500m		*Station A (1990–91) (70°53'S, 177°21'E) bottom depth: 1800 m*		*Station 287 GEOSECS (1972) (69°05'S, 173°30'W) bottom depth: 4149 m*		*Station 290 GEOSECS (1972) (58°00'S, 174°00'W) bottom depth: 5177 m*	
depth (m)	*^{226}Ra*	*depth (m)*	*^{226}Ra*	*depth (m)*	*^{226}Ra*	*depth (m)*	*^{226}Ra*
3	2.40 ± 0.11	20	2.37 ± 0.25	*5*	*2.83 ± 0.10*	*10*	*2.55 ± 0.10*
100	2.55 ± 0.31	80	2.59 ± 0.30	*66*	*2.90 ± 0.10*	*125*	*2.43 ± 0.14*
200	2.49 ± 0.32	150	2.58 ± 0.29	*183*	*2.81 ± 0.10*	*200*	*2.17 ± 0.10*
400	2.64 ± 0.31	300	2.77 ± 0.32	*318*	*2.90 ± 0.10*	*321*	*2.58 ± 0.10*
600	2.97 ± 0.32	500	2.85 ± 0.33	*537*	*3.25 ± 0.10*	*500*	*2.68 ± 0.10*
800	3.00 ± 0.47	700	3.01 ± 0.33	*780*	*3.43 ± 0.14*	*704*	*2.85 ± 0.14*
1000	3.77 ± 0.57	1000	3.46 ± 0.34	*1004*	*3.38 ± 0.14*	*905*	*2.90 ± 0.10*
		1500	3.73 ± 0.41	*1620*	*3.40 ± 0.10*	*1499*	*3.08 ± 0.14*
					(a)		(b)

(a) approximately constant values reported by the Authors to a depth of 3902 m.
(b) activity values steadily increase to a maximum of 3.76 Bq/m^3 at a depth of 4793 m.

from the same geographical area are included in Table 2. The values obtained, though limited in number, are nevertheless comparable to analogous literature data both for the Ross Sea[9] and for the Weddell Sea[13], collected during the GEOSECS program. It is possible to observe that the values reported range from 1.25 (Station 296, in proximity of New Zealand) to 3.64 (coastal station near the Italian base at Terra Nova Bay). All the other stations, located between 58° and 75° Lat.S, show homogeneous values. This can be attributed to the influence exerted by the Antarctic Convergence, which gives rise to an important oceanic frontal zone, due to the confluence of two different water masses: a) the warmer Circumpolar Bottom Water (forming in temperate areas) that reaches the surface in the circumpolar zone, where it cools and thereafter sinks; b) the colder Antarctic Surface Water (wind-driven to the north) which in proximity of Circumpolar bottom water sinks[20].

The Antarctic Convergence, limiting the mixing of the cold water masses it includes and the external temperate ones, makes the Antarctic continent practically "isolated" allowing for just a very slow and gradual exchange of heat, oxygen and carbon dioxide between the water masses in contact. The ^{226}Ra contents, which inside the Antarctic Convergence is higher due to the vertical mixing of water masses causing bottom resuspension of the parent ^{230}Th, are in accordance with this.

Data reported in Table 3 reflect what has been mentioned above. Variations along the depth profile reflect the combined effect of deep waters upwelling and mixing with southern waters.

It is to note that surface samples from the 1989–90 expedition available in much larger volumes, showed a correspondingly lower measurement uncertainty.

The higher activity values observed for stations 2SW and 7SW (in proximity of the Italian Base at Terra Nova Bay, 5 km off the coast line) can be attributed to the relatively shallow depth of the sea bottom, where sediment contribution of ^{226}Ra by resuspension can not be excluded. Moreover, a terrigenous contribution from the coast through erosional processes can result significant to the local ^{226}Ra activity observed.

A strong influence of the proximity of the seabed seems to be related as well to the data obtained from the vertical profiles at Station 10 and Station A with a bottom depth of 1500 m and 1800 m, respectively. The increase of ^{226}Ra activity with depth brings to the same higher levels observed in the surface samples previously mentioned, when approaching the bottom. This trend was observed also by Ku *et al.*[9] at the same location, further confirming the importance of bottom sediments influence on ^{226}Ra vertical distribution.

CONCLUSIONS

Data of ^{226}Ra preliminarly obtained from seawater samples collected in the Ross Sea, Pacific sector of the Antarctic Ocean, in past expeditions have allowed to elucidate two different aspects of the problems inherent to the use of ^{222}Rn as a tracer of gas exchange at the air-ocean interface: a) the development of an experimental procedure reliable and sensitive even during an oceanographic cruise, when ^{222}Rn is to be determined; b) the collection of basic information on the expected values of activities of the couple ^{226}Ra/^{222}Rn in the Antarctic marine environment.

As regards point a), it is to remark that although the apparatus so far used does not provide the possibility of instrumental and computational α/β discrimination, nevertheless, the results obtained demonstrate its reliability; in addition it has the suitable characteristics required for on-board operations, such as robustness and compactness. The adoption of resin adsorption of ^{226}Ra step for its preconcentration, though extensively used for the separation of other radionuclides from seawater, is a fairly innovative technique for ^{226}Ra itself, leading to remarkable analytical advantages, such as a higher precision and transport of less bulky sample load. In addition, the combination of recent scintillation counting improvements, allowing for α/β discrimination with a correspondingly lower detection limit (the LDA is in fact far lower even than ^{226}Ra activity in the Mediterranean seawater), further encourages the use of "radon deficit" method in the study of air-sea gas exchange and which is our primary objective as concerns our involvement in the National Program of Research in Antarctica.

Moreover, the uncertainty associated to measurement is greatly reduced when analyses are carried out on board, in the open sea, due to negligible instrumental background far off the high contribution to natural radioactivity of crustal rocks. Obviously, the need for on-board operations is dictated by the short half-life of ^{222}Rn, which is directly involved in the gas-exchange process and that, therefore, requires to be determined at the time of sampling.

As for point b), preliminary, data of ^{226}Ra obtained will be useful to locate the most suitable sites for seawater samplings in the next expeditions: special care will have to be paid where sea bottom is shallow, where sediment contribution to ^{226}Ra/^{222}Rn in the water column could significantly affect a correct deficit evaluation.

In addition, since the application of this method requires to know the vertical distribution of ^{226}Ra activity, it is to emphasize that these data independently can provide useful information, when used as water masses tracer. As reported by Ku *et al.*[9], it is possible, therefore, on the basis of different specific activities of ^{226}Ra (Bq/m^3), to distinguish among the main four ocean water masses circulating in the subantarctic oceans: a) surface water (1.3 ÷ 3.0); b) intermediate Antarctic water (2.3); c) intermediate circumpolar water (3.0); deep water (3.3 ÷ 3.8).

References

1. W. M. Smethie, Jr, T. Takahashi and D. W. Chipman, *J. Geophys. Res.*, **20**, 7005–7022 (1985).
2. P. P. Murphy, R. A. Feely, R. H. Gammon, D. E. Harrison, K. C. Kelly and L. S. Waterman, *J. Geophys. Res.*, **15**, 455–465 (1991).
3. T. H. Peng., T. Takahashi and W. S. Broecker, *J. Geophys. Res.*, **79**, 1772–1780 (1974).
4. Peng, W. S. Broeker, G. G. Mathieu and Y-H Li, *J. Geophys. Res.*, **84**(C5), 2471–2486 (1979).
5. C. Duenas, M. C. Fernandez and M. Perez Martinez *J. Geophys. Res.*, **88**(C3), 8613–8616 (1983).
6. P. S. Liss, *Phil. Trans. R. Soc. Lond.*, **A 325**, 93–103 (1988).
7. F. F. Koczy, *Proc. Second U.N. Int. Conf. Peaceful Uses of Atomic Energy*, Geneva, **18**, 351 (1958).
8. Y. C. Chung and H. Craig, *Earth Planet. Sci. Letters*, **17**, 306–318 (1973).
9. T.-L. Ku and M.-C. Lin, *Earth Plan. Sci. Lett.*, **32**, 236–248 (1976).
10. L. H. Chan, J. M. Edmond, R. F. Stallard, W. S. Broeker, Y. C. Chung, R. F. Weiss and T. L. Ku, *Earth Plan. Sci. Lett.*, **32**, 258–267 (1976).
11. Y. C. Chung, *Earth Plan. Sci. Lett.* **32**, 249–257 (1976).
12. W. S. Broeker, J. Goddard and J. L. Sarmiento, *Earth Plan. Sci. Lett.*, **32**, 220–235 (1976).
13. Y. C. Chung and M. D. Applequist, *Earth. Plan. Sc. Lett.*, **49**, 401–410 (1980).
14. W. S. Moore and J. Dymond, *Earth Plan. Sci. Lett.*, **107**, 55–68 (1991).
15. Scientific Committee on Antarctic Research, *SCAR Steering Committee for IGBP, Cambridge,* August 1992.
16. G. Queirazza, M. Roveri, R. Delfanti and C. Papucci, *Radionuclides in the study of marine processes* (P. J. Kershaw and D. S. Woodhead eds., Elsevier, 1991) pp. 94–104.
17. L. Cantelli, R. Delfanti, A. Moretti and M. Roveri, *Proceedings AIRP Conference on "Environmental radioactivity in the area of the Mediterranean Sea,* (C. Trivlzi and F. Nonnis Marzano eds, Univ. of Parma, 1994) pp. 205–211.
18. H. M. Prichard and K. Marien, *Anal. Chem.*, **55**, 155–157 (1983).
19. K. B. Rosson and R. J. Cantrell, *Health Phys.*, **59**, 125–131 (1990).
20. A. L. Gordon and J. A. T. Bye, J. *Geophys. Res.*, **77**, 5993–5999 (1972).

DISTRIBUTION OF MACRO- AND MICRO-COMPONENTS IN THE WATER COLUMN OF THE ANTARCTIC ROSS SEA AND IN SURFACE ANTARCTIC SNOW

P. PAPOFF, F. BOCCI and M. ONOR

Analytical Chemistry Section, Dept. of Chemistry, University of Pisa, Via Risorgimento 35, 56126 Pisa, Italy

Sample of Antarctic Ross Sea water were collected during the 1990/91 Italian Antarctic expedition: near the coast (74°40' S, 164°07' E) both with and without pack ice above the water column, and in the open ocean (70°53' S, 177°21' E) at different depths. In addition to alkaline and alkaline-earth elements, Se(IV), total inorganic Se, V, Cd, Pb and Cu were determined along the water column depth.

In the open sea the concentrations of the macro components increased by ca, 3% from 0 to 500 m in depth and then decreased lightly (1%) for all these analytes from 500 to 1500 m. Both soluble selenium (IV) and total inorganic Se, increased with increasing depth whereas the concentration of V was fairly constant when normalized to the sodium concentration of V. The dilution effect due to the pack ice under the water column could no longer be measured at a depth of 25 m.

Surface snow samples collected during the same expedition at varying distances from the sea, showed that the ionic concentration ratios relating to sodium are from two to five times higher than in sea water for potassium, two to four times for calcium, six to ten for magnesium, about 10^4 for lead and cadmium, and 10^5 for copper. In particular, it was found that alkaline and alkaline-earth elements exhibited the highest concentrations at a station 10 Km from the sea (100 m a.s.l.) and the lowest ones at a station 60 Km from the sea (700 m a.s.l.). No precise correlation was generally found among the normalized concentrations of heavy metals at the various sampling stations. For instance, the highest concentrations of lead, cadmium and copper were found at the last station, they were about 2.5 or 9 times higher than in the former, according to whether absolute or relative (to sodium) concentrations were considered.

KEY WORDS: Antarctica, sea water, snow, metals.

INTRODUCTION

This work is part of a series of research activities carried out in Terra Nova Bay with the aim of describing and characterising this Antarctic area from the environmental chemistry standpoint.

Two different matrices were considered. Firstly, sea water, to characterize the water column in terms of macro- and micro- components, in connection with the distance from the sea and the presence of pack ice over the sampling place. Secondly, surface, snow, with the target of further supporting the hypothesis that sea water aerosols are not the main factors causing the chemical content of snow.

As the samples were collected during the summer, when it rarely snows, the salt content in such samples refers to a larger exposition time to the atmosphere than for the deeper layers relating to the winter. All the findings however relate to the summer climate.

Since it is assumed that the chemical content in snow is not as homogeneously distributed as it is in sea water samples, the relative concentrations of some components (normalized to sodium concentration) are expected to play a very important role in data interpretation in addition to absolute ones.

Following the above considerations, samples of surface snow collected in Antarctica during the 1991/1992 Italian expeditions were analysed for Na, K, Ca, Mg and NH_4 in addition to Cd, Pb and Cu.

EXPERIMENTAL

Instrumentation. A Dionex model DX300 Ionic Chromatography (Dionex, Sunnyvale, CA, USA) equipped with an injection valve model 9126 Reodyne; a Pulsed Electrochemical Detector (PED) working in conductivity mode; a Dionex Micro Membrane Suppressor (CMMS II: 4 mm for sea water samples; 2 mm for snow samples) were employed. The separations were carried out on an Ionpac CS11 (250 mm × 2 mm i.d.) for snow samples, and on an Ionpac CG12 (50 mm × 4 mm i.d.) and and Ionpac CS12 (250 mm × 4 mm i.d.) for sea water samples. The loop volume was 50 μL. An AutoIon 450 for data acquisition and processing was used.

A DQP-1 pump (Dionex, Sunnyvale, CA, USA) was used for the preconcentration of snow samples (2 ml of samples) on an Ionpac CG11 (50 mm × 2 mm i.d.), and an AutoIon 450 for data acquisition and processing. Different cation exchange and C18 columns were used according to the solutes to be eluted.

A PAR MOD. 174 polarographic analyser, equipped with an AutoIon 450 for data acquisition and processing, was employed in the differential pulse mode. The electrochemical cell was equipped with a platinum wire counter-electrode and saturated calomel (SCE) and hanging drop mercury (HDME) as reference and working electrodes, respectively.

Reagents. All reagents were Suprapure grade materials. All standards, samples and reagents were prepared and stored in pre-cleaned polyethylene containers conditioned according to a procedure for trace element determination described elsewhere[1].

Sampling stations and sampling techniques

Sea water. Sea water samples were collected during the 1990–91 expedition by a go-flow system with teflonated bottles. Samples were taken in two sites (Figure 1) in order to obtain an oceanic profile (Station A: 70°53' S, 177°21' E) and a coastal one (Station B-Gerlache Inlet: 74°40' S, 164°07' E). Two series of samples were collected at Station B: before pack ice melting, through a hole in the ice, and after pack ice melting.

Immediately after sampling, sea water samples were filtered through a 0.45 μm pore size membrane filter, acidified at pH 2 and stored at 4°C. Cleaned and conditioned high-density polyethylene containers were used.

Snow. Surface snow samples were collected during the 1991–92 expedition in three sites (Figure 2): Priestley Glacier (station 03/SN/SUP/A/: 74°08' S, 162°48' E, about 700 m a.s.l.); Snowy Point (station 11/SN/SUP/A: 74°04' S, 165°19' E, about 100 m a.s.l.) and Kay Island (station 14/SN/SUP/A: 74°38' S, 163°45' E, about 100 m a.s.l.). After discarding about the first 2 cm layer, the snow samples were collected manually at

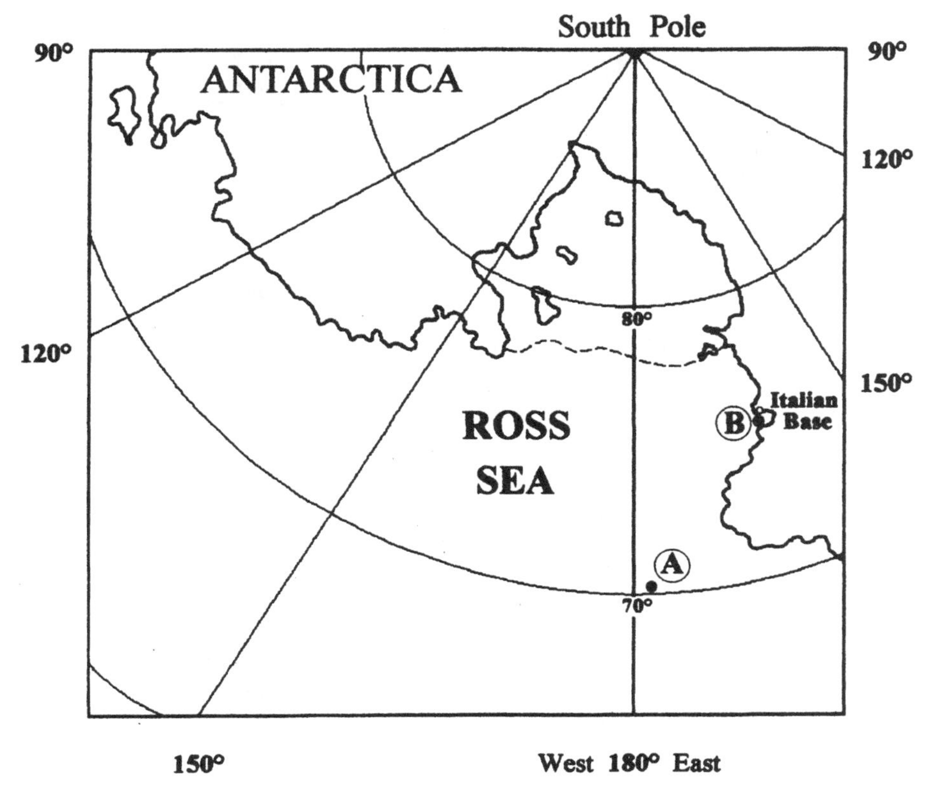

Figure 1 Location of sampling stations of sea water samples collected during the 1990/91 Italian expedition in Antarctica: **A** off shore and **B** in shore sampling stations. In station **B** (Gerlache Inlet) samples were collected before and after ice pack melting.

each station for a maximum depth of about 5 cm and a total volume of ca. 1 litre using a precleaned teflon scraper and precleaned sample wide-necked containers (1000 ml, high density polyethylene).

To prevent contamination the sampling procedure was always carried out against wind direction, around 500 m from the transportation means. All samples were stored at –20°C and analysed in our laboratory in Italy immediately after melting.

Procedures

Alkaline and alkaline earth elements

a) *Sea water.* Using an ion chromatographic (I.C.) procedure previously described[2], lithium, sodium, potassium, magnesium and calcium were determined on the same run after a 50-fold dilution of the sample: a 15 mM HCl was used in the isocratic elution mode on a CS12 Dionex column (250 mm × 4 mm i.d.).

Concentrations were estimated on three replicates by using for each set of measurements an external three point calibration plot. Concentrations of the analytes in

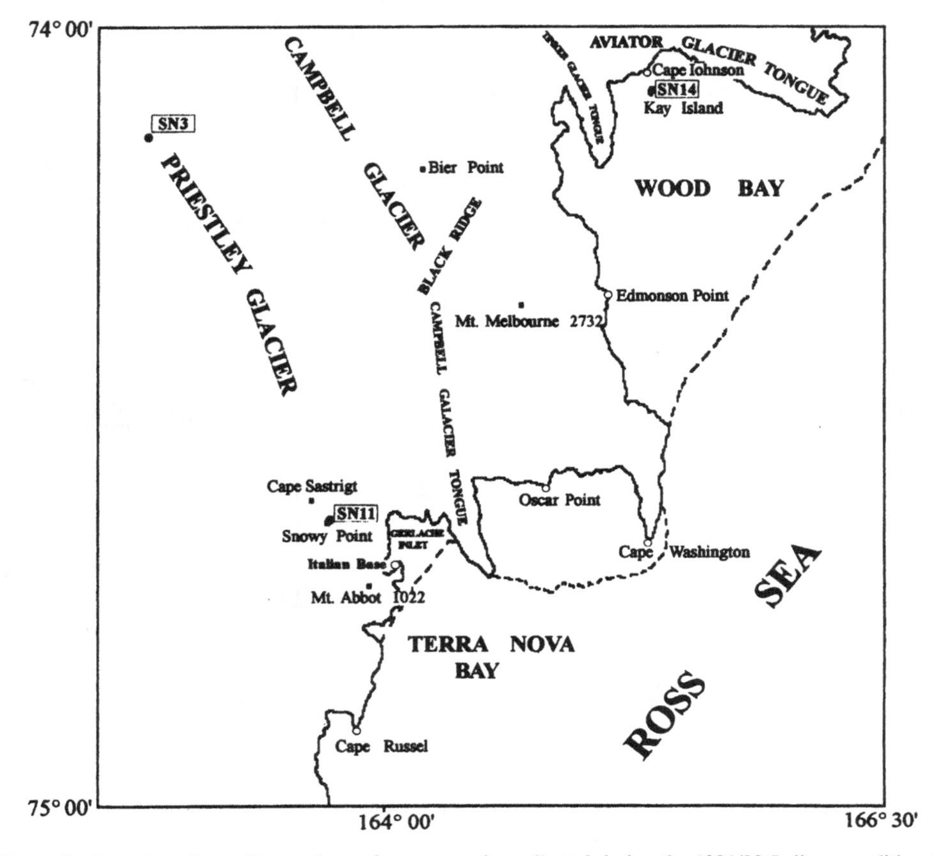

Figure 2 Location of sampling stations of snow samples collected during the 1991/92 Italian expedition in Antarctica.

the standard solution were arranged so that the signal of the sample for each analyte was around the middle of the plot. The concentrations, % relative confidence intervals (% R.I.C.), as obtained by regression analysis at a confidence level of $\alpha = 0.05$, around the mean value, were: Li: 4.6–5.4, Na: 0.03, K: 0.6–0.7, Mg: 0.3, Ca: 0.5.

b) *Snow.* After a preconcentration step on a CG11 Dionex Column[2], a 40 mM HCl mixed with 2 mM 2,3-Diamminopropionic monohydrochloride acid (DAP-HCl, NovaChimica, MI) eluent solution and a 40 mM HCl mixed with 8 mM DAP-HCl solution were used in a step gradient mode. The step gradient elution was changed after 5 minutes.

Blank contributions, due to release from the walls of the containers or from the laboratory environment, were made insignificant by making the whole chromatographic chain operate in an air-tight closed loop. Before each set of measurements a few runs with pure water were performed to wash out the system. Blank concentrations (nM) were: Na: 5, NH_4: 3, K: 0.3, Mg: 0.08, Ca: 0.5. Concentrations were estimated by regression analysis using, as for sea water, an external three point calibration plot.

Total soluble lead, cadmium and copper

These elements were determined both in sea water and snow by differential pulse anodic stripping voltammetry (DPASV), after UV decomposition of the organic matter present in the sample[3]. The estimated d.l. were: $2.7 \cdot 10^{-2}$ nM (3 ng.l^{-1}) for Cd, $2.9 \cdot 10^{-2}$ nM (6 ng.l^{-1})for Pb and $3.1 \cdot 10^{-1}$ nM (20 ng.l^{-1}) for Cu. The blank contributions due to reagents were typically below d.l. for all the elements considered.

In order to quantify the extent of sample-container exchanges, samples of pure water were analyzed for the above elements before and after their trip from our laboratory in Pisa to the Antarctic Italian Base and return.

The mean values together with their R.I.C.s ($\alpha = 0.05$), before and after the trip, were: 10.1 ± 16.5% and 11.5 ± 13.6% ng·l^{-1} for Pb, 29.3 ± 18.7% and 30.4 ± 20.1% for Cu. The cadmium content was always found far below the detection limit. The % r.s.d. values relevant to subsamples or to parallel samples were found to coincide and equal to: Cd 3%, Pb 4.5% and Cu 6%. Mean concentration values were estimated by regression analysis using a two-point standard addition method.

Se(IV) and total inorganic Se in sea water

Se(IV) was determined by differential pulse cathodic stripping voltammetry (DPCSV) according to the procedure initially proposed by Baltensperger[4]. Since the organic content in the sample affects the sensitivity of the method, ultraviolet irradiation of the sample, at pH 2 in the presence of 50 µl H_2O_2 for 60 ml sample, was found useful to perform before analysis. In these conditions we found that H_2O_2 decomposes organic substances such as humic acids or benzene without any contribution of organic selenium to Se(IV). Total inorganic Se was determined as Se(IV) according to the procedure[4,5], after UV reduction of Se(VI) to Se(IV) at pH 10. A two point standard addition procedure was used for concentration estimation. Mean concentrations and relevant % R.I.C. values are shown in Table 2. No blank contribution was found at the Se concentration level measured.

Vanadium

Based on reported procedures[6-8], vanadium, as V(IV + V), was determined by RP-HPLC after its derivatization in the sample by 5-Br-PADAP and H_2O_2, in the presence of Triton X-100, followed by its preconcentration on a C18 chromatographic pre-column. The elution was performed on an analytical Alltima-C18 5 µm Alltech column (250 mm length, 4.6 mm diameter) using a two-step eluent programme.

The peak area was found to vary linearly both with the preconcentration volume and the original concentration in the sample. Blank levels (related to the original samples) were lower than 20 ng.l^{-1} (0.4 nM). The usable linear dynamic range was from 0.8 to 200 nM. Mean values and relevant % R.I.C.s are shown in Table 2.

RESULTS AND DISCUSSION

Sea water. The concentration values for some alkaline and alkaline-earth elements present in the sea water samples are reported in Table 1. This shows of each element: the

Table 1 Ice pack and water column depth effects on alkaline and alkaline-earth element ion concentrations in Antarctic sea water samples gathered at stations A and B (see Figure 1). $\overline{X}_i$: mean concentration (mM) (on three replicates) estimated by regression analysis using an external calibration curve (see text) daily controlled. $R_{i,z} = (\overline{X}_i/\overline{X}_{Na})_z$: normalized concentration of the i analyte, referred to sodium concentration, both measured at the same station Z (R values obtained by the original not rounded untrounched data of $\overline{X}_i$. For an estimation of % R.I.C. ($\alpha = 0.05$) see text.

	$\overline{X}_i$					$(\overline{X}/\overline{X}_{Na})_z = R_{i,z}$			
Sample name	*Li*	*Na*	*K*	*Mg*	*Ca*	*[Li$_i$]/[Na$_i$]*	*[K$_i$]/[Na$_i$]*	*[Mg$_i$]/[Na$_i$]*	*[Ca$_i$]/[Na$_i$]*
SWA-20	$22.6 \cdot 10^{-3}$	461	9.90	52.2	10.2	$4.9 \cdot 10^{-5}$	0.0214	0.113	0.0221
SWA-150	$22.7 \cdot 10^{-3}$	470	10.1	53.5	10.3	$4.8 \cdot 10^{-5}$	0.0214	0.114	0.0220
SWA-500	$22.7 \cdot 10^{-3}$	473	10.1	53.8	10.4	$4.8 \cdot 10^{-5}$	0.0214	0.114	0.0220
SWA-700	$25.5 \cdot 10^{-3}$	470	10.0	53.3	10.3	$5.4 \cdot 10^{-5}$	0.0213	0.114	0.0220
SWA-1500	$24.3 \cdot 10^{-3}$	468	10.0	53.3	10.3	$5.2 \cdot 10^{-5}$	0.0214	0.114	0.0221
SWB8-0.5	$21.5 \cdot 10^{-3}$	461	9.90	51.6	10.0	$4.7 \cdot 10^{-5}$	0.0215	0.112	0.0218
SWB8-25	$17.8 \cdot 10^{-3}$	464	9.86	52.0	10.1	$3.8 \cdot 10^{-5}$	0.0213	0.112	0.0217
SWB8-250	$16.5 \cdot 10^{-3}$	462	9.94	51.7	10.1	$3.6 \cdot 10^{-5}$	0.0215	0.112	0.0218
PWB6-0.5	$14.7 \cdot 10^{-3}$	421	9.22	48.1	9.30	$3.5 \cdot 10^{-5}$	0.0219	0.114	0.0221
PWB6-25	$24.7 \cdot 10^{-3}$	465	10.1	53.8	10.4	$5.3 \cdot 10^{-5}$	0.0217	0.116	0.0224
PWB6-250	$26.6 \cdot 10^{-3}$	468	10.2	54.4	10.5	$5.7 \cdot 10^{-5}$	0.0218	0.116	0.0223

individual $\overline{X}_i$ mean concentration values (three replicates) for the samples taken at various depths, z; the $R_{i,z} = (\overline{X}_i/\overline{X}_{Na})_Z$ values, i.e. the normalized concentrations of any element related to the concentration of sodium at the same z.

From these data, it was found for all the i solutes, with the exception of lithium, that:

i) in the open ocean the absolute concentrations $\overline{X}_i$ increased from 0 to 500 m; and then decreased to 1500 m, although not significantly from hydrographic view point: owing to the accuracy and precision levels of the chromatographic procedure used it was possible to quantify a decrease in concentration, for all the elements considered, as low as 1%.

ii) near the coast (Station B) the effect of the ice pack under the sea water corresponds to a dilution of sea water of about 7% at 0.5 m under the surface of ice-sea water, which is no longer evident at 25 m.

The reproducibility of R_is among depth for each element is a direct, experimental assessment of the accuracy of the analytical procedures used.

The fact that lithium shows a particular trend, can only be partially explained in terms of a lower accuracy in its determination owing to its low level of concentration and to the fact that its chromatographic peak is adjacent to that of sodium with a concentration ratio of $1{:}10^5$. The erratic behaviour of lithium is supported by the data of concentrations measured at the same station before and after pack ice melting. The lithium concentration increases or decreases with depth, depending on whether the samples were collected before or after ice pack melting, respectively, while the $[Li]_i/[Na]_i$ ratio reaches the highest values at 250 m in depth before melting and 0.5 m after pack ice melting. Interaction with particulate and sediments is considered likely[9].

Accurate measurements of salinity, performed at the same conditions by Innamorati *et al.*[10], are consistent with our findings.

As for the micro-elements V and Se in sea water (see Table 2), in spite of their concentration at nM level, the following findings were obtained:

i) vanadium showed a fairly constant ionic concentration ratio with sodium $(6.7 \pm 0.6) \cdot 10^{-8}$ for all the considered stations. The range of V concentrations observed in

Table 2 Effects of water column depth on ion concentration (nM) of some microcomponents in Antarctica sea water samples collected at stations A and B (see Figure 1). In the case of V and total inorganic Se, measurements were also performed at 0.5 m below sea water level before and after ice pack melting.
% R.I.C. (a = 0.05) values were estimated by regression analysis using two spikes for V and two standard additions for all the other components. Three replicates for each sample.

Sample name	*V*	*R.I.C.%*	*Tot. inorganic Se*	*R.I.C.%*	*Se(IV)*	*R.I.C.%*	*Cu*	*R.I.C.%*	*Pb*	*R.I.C.%*	*Cd*	*R.I.C.%*
SWB8-0.5 A	34.9	1.09	1.11	9.78	–	–	–	–	–	–	–	–
SWB8-0.5 B	–	–	1.00	12.3	–	–	–	–	–	–	–	–
PWB6-0.5 A	27.4	2.91	0.62	15.3	–	–	–	–	–	–	–	–
PWB6-0.5 B	–	–	0.67	16.3	–	–	–	–	–	–	–	–
SWA-20	30.9	0.82	0.95	6.28	0.067	32.5	2.69	14.3	0.84	11.2	0.63	12.8
SWA-700	32.1	1.27	1.64	4.15	0.39	14.5	3.71	17.7	0.97	6.3	0.72	5.7
SWA-1500	27.9	1.65	2.46	2.83	0.48	4.25	3.83	12.2	0.67	10.9	0.59	17.1

Antarctica sea water was comparable with that observed by J. H. Martin[9] in the eastern North Pacific. According to the findings of Morris[11], the concentration of V in oceans is 23 ± 3 nM with no discernible depth-related patterns.

ii) selenium, both as Se(IV) and total inorganic Se, presents an increasing concentration at an increasing depth of the water column, with a trend similar to that observed by Measures *et al.*[12] in the open ocean.

Snow. The chemical content in surface snow samples depends upon many variables which are mutually correlated in a complicated way. The following play an important role:

i) the time of sampling with respect to the onset of precipitation and to the cyclonic storm period or season[13,14];

ii) the scavenging ability[15,16] of the snow with respect to the air;

iii) the exposure time to sunlight between snowing and sampling[17];

iv) the layer of surface snow considered and the type of precipitation such as hoarfrost or snow[17,18]; the distance and the elevation of the sampling point from the sea[19].

In this sense the chemical content of surface snow is expected to change, even significantly, in the same geographical area according to the time, location and depth of the sampling. As a general rule, the determination of traces in snow, such as alkaline and alkaline earth elements down to 10 nM, does not involve serious analytical problems in terms of the accuracy of the estimated mean values. It is in the case of ultratraces, such as heavy metals in snow, where concentrations of the order of 1 nM or less are involved, that there is a considerable probability that one or more samples become inadvertently contaminated during the handling procedure. As a matter of fact, and this is a fundamental preliminary condition, before assigning a precise meaning to the observed variations in the chemical content of snow samples, sound information must be acquired about the overall reproducibility of the procedure steps and on the relevant accuracy, at real concentration levels in the sample.

The literature and our own direct experience show that the occurrence of outliers is always possible, at least at a $pg \cdot g^{-1}$ level, no matter how good the procedures used from sampling to analysis are. Reducing the probability of using outliers is thus the primary goal: this can be achieved by considering many samples collected in the same area or many parallel samples of the same origin collected in few places.

Two enlightening examples will now be discussed concerning the results obtained in Antarctica from two different groups of researchers who are, in our opinion, among the best in obtaining the most accurate data currently possible.

Boutron and Patterson[20] were able, in the case of lead in ice core, to recognize occasional contamination by the profile at each depth of the considered analyte (lead) concentration vs the distance from the centre of the core. In spite of all the care taken, the dependence of lead concentration on the age of the core related to this century has given rise to totally different interpretations according to whether the data related to 1900–1950 or 1950–1980 are considered to be more reliable (max. difference 4 $pg \cdot g^{-1}$).

Obviously seasonal effects and any short-term changes which may affect surface snow, are no longer evident in core samples in which each sample refers to several years of snow accumulation.

In the case of surface snow, Völkening and Heumann[17], measured several elements on the same Antarctic field using DPASV, whenever possible, and IDMS (Isotopic Dilution Mass Spectrometry). In the case of lead, they found a reproducibility among portions of the same sample within 10% and among parallel samples within 30%, in the DPASV measurements. Considering that the lead concentration was in the range 5–120 $pg \cdot g^{-1}$ this reproducibility is very good. When the results obtained by DPASV and by IDMS are compared, some significant differences can be observed for the same samples, apart from the effect of snow aging. For instance, in one case they found 120 $pg \cdot g^{-1}$ using DPASV and 178 with IDMS, in another case 4.5 $pg \cdot g^{-1}$ with DPASV and 25.8 with IDMS.

As for Cd, some samples with a concentration of ca. 4 $pg \cdot g^{-1}$, as found with IDMS, were always under d. l. using DPASV, while for other samples at ca. 1 $pg \cdot g^{-1}$ level the two procedures substantially agreed. In addition, the authors disregarded 20% of the results related to the higher concentrations because they found that about 20% of all blank determinations carried out in the Antarctic laboratory, were considerably higher than the mean of blanks due to contamination problems.

Their findings clearly show that the combined effects of sampling date (within one month) and of the sampling origin (within a field of 700 × 700 m^2) may lead to snow content variations of 3 to 15 times depending on the element considered. In particular, the observed concentration range ($pg \cdot g^{-1}$) was 3–40 for lead 0.2–3 for Cd and 11–30 for Cu in surface snow, with an enrichment factor of 3–15 when hoar-frost samples are considered, compared to snow in the same location.

From all the above considerations, some important conclusions can be drawn:

i) no matter how good the sampling and analytical procedures used so far are (see f.i. ref. 21), measured concentrations in snow up to 5–8 $pg \cdot g^{-1}$ may contain an unknown amount of bias (provided that the estimation of the mean is statistically correct). Only in the case of cross-core profiles[20] can the effect of sampling contaminations be evidenced; while the other effects including analytical procedure bias remain unknown. Since the latter contributions may not be reproducible among samples, also correlations among data obtained from the same source are uncertain at this $pg \cdot g^{-1}$ level.

ii) with reference in particular to the findings[17,18] it is possible to dispute some trends whereby data sets at low $pg \cdot g^{-1}$ levels are considered to have been affected by contamination since they are five to ten times higher than other data obtained years before in another Antarctic region.

Any further research performed in different areas of Antarctica and extended over several years, will therefore lead to an improvement in understanding the dependence of the local effects on distance from the sea, elevation and incidental mountain protection against snow precipitation and wind, provided that the procedures used are carefully tested and found suitable.

If concentrations of several analytes are measured in surface or depth snow, additional information can be obtained on the origin of each analyte and, in particular, on whether they are to some extent generated by human activities.

It is generally accepted from the large amount of literature on this topic, that marine salt and atmospheric dust from local sources, only contribute a part of the ion chemistry of the snowfall, the greater part being contributed by gas derived aerosols.

A technique that has proved to be useful for a first step identification of the source of a given trace element *i*, is to normalize their concentrations with respect to a reference element. Frequently, Na is used as an indicator of atmospheric sea salt and Al as an indicator of weathered crustal material. The enrichment factor related to sea water is defined by:

$$EF_{sea,i} = ([I]/[Na])_{sample}/([I]/[Na])_{sea\ water}$$

and likewise when the EF is related to crustal material, the reference element is Al[22].

Under a large sea-air exchange program concerning the aerosol composition at the Enewetak Atoll (Pacific Ocean), Arimoto and Duce[15] were able to distinguish three groups of elements on the basis of their R_{sea} and $R_{crust.}$ values. Elements such as Br, Cl, Mg, K, Ca, Rb, were found to originate from the sea (R_{sea} were near 1); a large group, including Al, Fe, Co, Cr and Cu, were found to have crustal origin ($R_{crust.} < 3–5$); and elements such as Zn, Pb, Cd, Se and I were enriched significantly or predominantly controlled by other influences: an enrichment step was assumed to take place during the period spent in air.

Substantially similar conclusions have recently been reached by other authors[18,23,24] using Fe as a reference element, for the Atlantic Ocean and Antarctica eco-systems. In the case of Antarctica surface snow[18] the $EF_{crust.}$ was in the range: 100–820 for Cd, 32–97 for Pb, 8–20 for Cu and only 0.5–2 for Cr.

The interpretation of these different EF ranges was that the predominance of other influences besides crustal weathering sharply decreases from Cd to Cu, becoming zero for Cr (only terrestrial origin).

Table 3 shows that the highest concentrations of alkaline and alkaline-earth elements were found at station SN11 (100 m elevation, 10 Km from the sea, behind a 500 m hill), and the lowest ones at SN3 (700 m elevation, 60 Km from the sea). The EF inspection shows that none of the salt contents in the samples have a predominant marine origin.

It is worth comparing our normalized values with those calculated by Johnson and Chamberlain original data[25], relating to sampling in Law Dome (East Antarctica) performed in 1986. At 5 Km from the coast and 380 m elevation their EF's with respect to sea water were 1.3 for K, 1.0 for Mg and Ca, which are quite different from our data at SN11. In addition, the EF's for their sampling at 110 Km distance and 1400 m elevation (3.2 for K, 8.1 for Mg and 2.1 for Ca) were comparable with ours for SN14, this showing that the distance from the sea and elevation are not the only determinant factors[26].

The concentrations of heavy metals, were on average 2.5 higher at SN3 than at SN11 or SN14, while the normalized to sodium concentrations were about 10 times higher at SN3. This fact can be interpreted by assuming that during snowing or later before sampling, the air from the interior of the continent, relatively enriched in heavy metals, was mixed with the coastal air, relatively enriched in sodium, in ways which may differ depending on the situation.

The fact that for each heavy metal considered, the concentrations are essentially constant at SN11 and SN14 and consistently lower than at SN3 (the furthest one from the Italian Station), may be seen as a proof that the contamination due to human activities in the base, is quite low, if any.

Table 4, which should not be seen as a comprehensive summary of literature data, shows the interval of concentrations (nM) obtained by various authors, along with information about the depth of the snow layer considered and the concentrations in sea water. For further data, concerning snow and ice see references 17 and 20.

Apart from Landy[27], whose concentration values appear to be quite high, Table 4 highlights that bearing in mind the unpredictable effects of different Antarctic regions and sampling data as well, the ionic concentrations of heavy metals generally increase as both the distance from the surface and the thickness of the snow layer collected decrease. However, they may vary by a factor of ten or more depending on the sampling location, even when a restricted field is considered[17].

Normalized concentrations (with respect to sodium) show no correlation among different sampling sites, thus confirming that sea water aerosol is not the direct cause of heavy metal content in snow. Also, neither the distance from the sea nor the elevation are determinant in sea salt content in surface snow.

Table 3 $\overline{X}_i$ mean concentrations and $R_{i,z}$ normalized concentrations on three replicates of some elements in snow samples gathered (see Figure 2). The EF enrichment factors refer to sea water. (a) and (b) refer to I.C. and DPASV measurements respectively. For an estimation of % R.I.C. ($\alpha = 0.05$) see text.

	Element							
Station i	*[NH4]*[a]	*[Na]*[a]	*[K]*[a]	*[Mg]*[a]	*[Ca]*[a]	*[Pb]*[b]	*[Cu]*[b]	*[Cd]*[b]
SN3	0.30 (7)	1.64 (14)	0.17 (14)	0.19 (11)	0.06 (15)	$2.7 \cdot 10^{-4}$ (9)	$3.6 \cdot 10^{-3}$ (15)	$3.9 \cdot 10^{-4}$ (7)
SN11	0.70 (2)	8.35 (4)	0.39 (7)	0.62 (5)	0.71 (9)	$1.1 \cdot 10^{-4}$ (12)	$1.9 \cdot 10^{-3}$ (17)	$1.2 \cdot 10^{-4}$ (8)
SN14	2.14 (3)	5.65 (2)	0.34 (4)	0.41 (20	0.25 (5)	$1.3 \cdot 10^{-4}$ (11)	$1.4 \cdot 10^{-3}$ (14)	$1.4 \cdot 10^{-4}$ (6)
	R_i							
Station i	R_{NH4}		R_K	R_{Mg}	R_{Ca}	R_{Pb}	R_{Cu}	R_{Cd}
SN3	$1.8 \cdot 10^{-1}$		$1.0 \cdot 10^{-1}$	$1.2 \cdot 10^{-1}$	$3.7 \cdot 10^{-2}$	$1.6 \cdot 10^{-4}$	$2.2 \cdot 10^{-3}$	$2.4 \cdot 10^{-4}$
SN11	$8.4 \cdot 10^{-2}$		$4.7 \cdot 10^{-2}$	$7.4 \cdot 10^{-2}$	$8.5 \cdot 10^{-2}$	$1.3 \cdot 10^{-5}$	$2.3 \cdot 10^{-4}$	$1.4 \cdot 10^{-5}$
SN14	$3.8 \cdot 10^{-1}$		$6.0 \cdot 10^{-2}$	$7.3 \cdot 10^{-2}$	$4.4 \cdot 10^{-2}$	$2.3 \cdot 10^{-5}$	$2.5 \cdot 10^{-4}$	$2.5 \cdot 10^{-5}$
Sea water			$2.10 \cdot 10^{-2}$	$1.18 \cdot 10^{-2}$	$2.20 \cdot 10^{-2}$	$4.9 \cdot 10^{-10}$	$6.23 \cdot 10^{-9}$	$7.47 \cdot 10^{-10}$
	E.F.							
Station i								
SN3			4.9	9.8	1.7	$3.4 \cdot 10^{5}$	$3.5 \cdot 10^{5}$	$3.2 \cdot 10^{5}$
SN11			2.2	6.3	3.9	$2.7 \cdot 10^{4}$	$3.7 \cdot 10^{4}$	$1.9 \cdot 10^{4}$
SN14			2.9	6.1	2.0	$4.7 \cdot 10^{4}$	$4.0 \cdot 10^{4}$	$3.3 \cdot 10^{4}$

Table 4 Interval of concentration (nM) of Cd, Pb and Cu found in Antarctica snow or sea water by different authors, in different sites and years. For further data concerning snow and ice, see [20] and [17]. The depth column shows the depth intervals from the surface inside which the layers of snow were collected.

<table>
<tr><th rowspan="2">Ref.</th><th rowspan="2">Expedition</th><th rowspan="2">Sample type</th><th rowspan="2">Depth</th><th colspan="2">Cd</th><th colspan="2">Pb</th><th colspan="2">Cu</th></tr>
<tr><th>min</th><th>max</th><th>min</th><th>max</th><th>min</th><th>max</th></tr>
<tr><td>20</td><td></td><td>Snow core</td><td></td><td></td><td></td><td>0.005</td><td>0.03</td><td></td><td></td></tr>
<tr><td>27</td><td>1976</td><td>Snow</td><td>Depth
not specified</td><td colspan="2">3.3</td><td colspan="2">3.9</td><td></td><td></td></tr>
<tr><td rowspan="2">17</td><td rowspan="2">1987</td><td rowspan="2">Snow
Hoar-Frost</td><td rowspan="2">Surf.
(15–30 cm)</td><td>0.002</td><td>0.03</td><td>0.015</td><td>0.2</td><td>0.2</td><td>0.5</td></tr>
<tr><td colspan="2">0.13</td><td colspan="2">0.9</td><td></td><td></td></tr>
<tr><td>This work</td><td>1991/92</td><td>Snow</td><td>Surf.
(2–5 cm)</td><td>0.12</td><td>0.39</td><td>0.11</td><td>0.27</td><td>1.4</td><td>3.6</td></tr>
<tr><td>28</td><td>1990/91</td><td>Snow</td><td>Surf.
25 cm</td><td></td><td></td><td colspan="2">0.24
0.011</td><td></td><td></td></tr>
<tr><td>29</td><td>1988/89</td><td>Snow</td><td>Surf.
(2–5 cm)</td><td>0.10</td><td>0.27</td><td></td><td></td><td>0.71</td><td>1.42</td></tr>
<tr><td>30</td><td>1990/91</td><td>Flowing
Melt waters</td><td></td><td>0.05</td><td>0.21</td><td></td><td></td><td></td><td></td></tr>
<tr><td>31</td><td>1989/90</td><td>Sea water</td><td></td><td>0.13</td><td>0.28</td><td>0.07</td><td>0.14</td><td>1.49</td><td>2.22</td></tr>
</table>

Finally, absolute concentrations of some elements in Antarctic surface snow and sea water are comparable as shown in Table 4 for Cd, Pb and Cu. The same was found for Zn, Tl and Fe:

Sample type	Zn	Tl	Fe
Sea water	4[31]	0.06[9]	8[31]
Surface snow	0.5–7.6[17]	0.001[17]	9–27[17]
flow. melt. snow		0.05–0.2[30]	

CONCLUSIONS

In Antarctic sea water it was found that the total inorganic Se concentration increases with depth, while the V concentration remains fairly constant regardless of depth and the presence of ice pack when referred to the same sodium concentration. These trends are similar to those found in oceans.

Regarding snow, from an analysis of literature data and our findings, it may be concluded that the chemical content in Antarctica surface snow may vary significantly even when a restricted field and a limited time interval is considered for sampling. The element contents increase as the distance from the surface and the thickness of the snow layer collected decrease. Thus, whenever problems related to contamination of the sample are avoided, local and seasonal effects due to air-snow interaction are amplified when surface snow is considered.

Owing to the continuous improvements of sampling and analysis procedures, it is likely that the actual bias in the determination of element concentrations in snow, which we estimate to be nowadays of some $pg \cdot g^{-1}$ will further be reduced. Incidentally, it is worth observing that absolute concentrations of metals such as Cd, Pb, Cu, Tl, Zn and Fe, in Antarctic surface snow and sea water are comparable in spite of the completely different values of their concentration normalized to sodium.

Acknowledgements

The authors wish to express their sincere appreciation to the researchers of the "Impatto Ambientale—Metodologie Chimiche" team for their accurate and skilful work. The financial support from ENEA—Antarctic Project is also acknowledged.

References

1. M. Betti and P. Papoff, *CRC Crit. Rev. Anal. Chem.*, **19**, 271 (1988).
2. P. Papoff, M. Onor and M. Betti, *Intern. J. Environ. Anal. Chem.*, **55**, 149–164 (1994).
3. M. Betti, R. Fuoco and P. Papoff, *Ann. Chim. (Rome)*, **79**, 689–699 (1989).
4. U. Baltensperger and J. Hertz, *Anal. Chim. Acta*, **172**, 49–56 (1985).
5. C. I. Measures and J. D. Burton, *Anal. Chim. Acta*, **120**, 177–186 (1980).
6. Y. Zhao and C. Fu, *Anal. Chim. Acta*, **230**, 23–28 (1990).
7. J. Miura, *Anal. Chem.*, **62**, 1424–1428 (1990).
8. J. Miura, *Analyst*, **114**, 1323–1329 (1989).
9. K. W. Bruland, *Trace elements in sea water chemical oceanography*. Vol **8**, chap. 45 (Academic Press, London 1983).
10. M. Innamorati, L. Lazzara, G. Mori, C. Nuccio and V. Saggivo, *Nat. Sc. Com. Ant., Ocean. Camp.*, 1989–90, Data Rep., (1991) I: 141, 252.
11. A. W. Morris, *Deep-Sea Res.*, **22**, 49 (1975).
12. C. I. Measures, B. Grant, M. Khadem, D. S. Lee and J. M. Edmond, *Earth Planet. Sci. Lett.*, **71**, 1–12 (1984).
13. A. A. Matveev, *Meteorologocheskie Issledovania Sbornik Statei*, **5**, 100–107 (1963).
14. A. A. Matveev, *J. Geophys. Res.*, **75**, 3686–3690 (1970).
15. R. Arimoto and R. A. Duce, in: *Sources and Fates of Aquatic Pollutants* (R. A. Hites and S. J. Eisenreich eds., Advances in Chemistry Series n. 216 A.C.S., Washington, 1987), pp. 131–150.
16. D. M. Settle and C. C. Patterson, *J. Geophys. Res.*, **87**, 8857–8869 (1982).
17. J. Völkening and K. G. Heumann, *Fresenius Z. Anal. Chem.*, **331**, 174–181 (1988).
18. K. G. Heumann, *Anal. Chim. Acta*, **283**, 230–245 (1993).
19. I. B. Campbell and G. G. C. Claridge, in: *Antarctica: Soils, Weathering Processes and Environment.* (Elsevier eds., New York 1987) pp. 239–273.
20. C. F. Boutron and C. C. Patterson, *Geochim. Cosmoschim. Acta*, **47**, 1368 (1983).
21. E. W. Wolff and D. A. Peel, *Nature*, **313**, 535–540 (1985).
22. S. R. Taylor, *Geochim. Cosmochim. Acta*, **28**, 1273–1285 (1964).
23. J. Völkening and K. G. Heumann, *J. Geophys. Res.*, **95D**, 20623 (1990).
24. N. Radlein and K. G. Heumann, *Intern. J. Environ. Anal. Chem.*, **48**, 127 (1992).
25. B. B. Johnson and I. M. Chamberlain, *Geochim. Cosmochim. Acta*, **45**, 771–786 (1981).
26. M. R. Legrand and R. J. Delmas, *Ann. Glaciol.*, **10**, 1–5 (1988).
27. M. P. Landy, *Anal. Chim. Acta*, **121**, 39–49 (1980).
28. G. Scarponi, C. Barbante, P. Cescon, *Atti del 3° Convegno P.N.R.A.: Impattto Ambientale-Metodologie Chimiche Venezia 10–11, March, 1994.*
29. G. Saini, C. Baiocchi, D. Giacosa and M. A. Roggero, *Ann. Chim. (Rome)*, **81**, 317–324 (1991).
30. G. Saini, V. Pedrini, D. Giacosa and P. R. Trincherini, *Atti del 3° Convegno P.N.R.A.: Impatto Ambientale-Metodologie Chimiche Venezia 10–11, March*, 1994.
31. P. Cescon, R. Fuoco and P. Papoff, *Intern. J. Environ. Anal. Chem.*, **55**, 91–119 (1994).

RADIOACTIVE AND STABLE ISOTOPES IN ABIOTIC AND BIOTIC COMPONENTS OF ANTARCTIC ECOSYSTEMS SURROUNDING THE ITALIAN BASE

C. TRIULZI,[a] F. NONNIS MARZANO,[a] A. CASOLI,[b] A. MORI,[a] and M. VAGHI[a]

[a]Department of General Biology and Physiology. University of Parma, Viale delle Scienze, 43100 Parma, Italy. [b]Department of General and Inorganic Chemistry, Analytical Chemistry, Physical Chemistry, University of Parma, Viale delle Scienze, 43100 Parma, Italy

Results concerning the analysis of natural (K-40, Th-232, U-238) and anthropogenic (Sr–90, Cs-137, Pu-238, Pu-239, 240) radioactivity determined in samples collected during the PNRA (National Program for Antarctic Research) 1990–91 and 1991–92 Scientific Expeditions, are presented. The data refer to samples of the terrestrial, lacustrine and marine ecosystems surrounding the Italian Base in the Terra Nova Bay (Ross Sea) territory with special emphasis on the Cs-137 biogeochemical behaviour. In particular, the role of the organic substance in the radionuclide transfer has been evaluated through statistical correlation analysis between the Cs-137 concentrations and organic matter, organic carbon and nitrogen contents determined in samples of marine and lacustrine sediments.

KEY WORDS: Antarctica, Terra Nova bay, artificial radioisotopes, natural radioisotopes, radioecology.

INTRODUCTION

Natural and artificial radioactivity measurements have been executed during the last five years on environmental samples collected in the terrestrial, lacustrine and marine ecosystems surrounding the Italian Terranova Base in Antarctica.

The results concerning the PNRA (National Program for Antarctic Research) 1987–88, 1988–89 and 1989–90 Scientific Expeditions have already been reported in previous papers[1–4]. In this work the unpublished results obtained during the next 1990–91 and 1991–92 Campaigns are presented.

The major part of the work has been oriented to the evaluation of the Cs-137 and the natural K-40, Th-232, U-238 presence in different environmental components of the three ecosystems. Moreover, data of H–3, Pb–210 and Po–210 was obtained in the past in some ice and compact snow cores[5].

The biogeochemical behaviour of Cs-137 in the Antarctic environment is also discussed. In particular the role of the organic substance in the radionuclide transfer has been evaluated through statistical correlation analysis between the Cs-137 concentrations and organic matter, organic carbon and nitrogen contents determined in samples of marine and lacustrine sediments.

Table 1 Sampling stations of the 1990–91 and 1991–92 campaigns not reported in previous publications.

Code	*Sample*	*Latitude (S)*	*Longitude (E)*
Terrestrial ecosystem			
SN 17	Snow	74°43'	164°01'
VO 14 Lich.	Lichens	74°04'	165°19'
VO 14 MOSS	Moss	74°04'	165°19'
SL 8	Soil	74°20'	165°07'
SL 14	Soil	74°04'	165°19'
SL 20	Soil	74°43'	164°01'
Lacustrine ecosystem			
LW 17/3	Lake water	74°43'	164°01'
AG 8	Algae	74°20'	165°07'
AG 12	Algae	74°59'	162°33'
AG 17	Algae	74°43'	164°01'
LS 8	Sediment	74°20'	165°07'
LS 12	Sediment	74°59'	162°33'
LS 17	Sediment	74°43'	164°01'
LS 20A	Sediment core	74°43'	164°01'
LS 20B	Sediment core	74°43'	164°01'
Marine ecosystem			
PW 5	Pack water	74°41'20"	164°06'56"
PW 6	Pack water	74°38'04"	164°17'17"
PW 7	Pack water	74°40'05"	164°13'38"

Some preliminary results on Sr–90 and Pu-239 + 240 concentrations obtained by radiochemical separations carried out on terrestrial and lacustrine samples collected during the 1990–91 Expedition, are also presented. Such determinations were executed as part of a collaboration activity with Prof. C. Testa and Dr. C. Roselli of University of Urbino.

Furthermore, part of this research was oriented towards the determination of the earth-alkaline elements calcium, magnesium, barium and strontium, besides the contents of potassium and uranium in seawater samples collected in coastal stations.

All sampling stations refer to the classic Italian study area at Terra Nova Bay and its inlet as described in the 1990–91 and 1991–92 Expedition final reports[7] and in previous papers for the marine environment[8]. The list of sampling stations is provided in Table 1.

MATERIALS AND METHODS

Samples of superficial unfiltered sea and lacustrine water collected during the IVth Expedition (1990–91) were submitted to the NCFC extraction method. This technique allows higher Cs-137 radiochemical separation yields (almost 100%) compared to the ones obtained with the classical AMP method[1,2]. In this technique, the liquid sample (300–400 liters for sea water and 95 liters for lake water) is pumped through NCFC-bed columns (ammonium hexacyanocobalt (II) ferrate (II) supported on silica gel), equipped with a flux counter (constant flux 100 ml/min), whose property is capturing the cesium isotopes[6].

Lacustrine and marine sediments, soil, lichens and moss, lacustrine algae, ichthyofauna and macrofauna organisms were collected during the different scientific expeditions. The different matrices were oven-dried at 105°C, minced and then introduced into standard source containers to be counted for gamma spectrometry. For what concerns the marine organisms, biometric analysis, sex recognition and anatomical dissection were executed before the drying procedure.

The analytical source obtained were counted for gamma spectrometry by means of two PGT Silena high resolution germanium detectors (FWHM 1.8 KeV at 1.33 MeV) for quantitative evaluation of the gamma emitters. All instruments had been previously calibrated through an intercalibration exercise with the IAEA- Marine Environment Laboratory (Monaco).

The gamma energy considered ranged between 0.08 and 1.5 MeV with direct detection of Cs-137 (0.662 MeV) and K-40 (1.461 MeV) besides determination of Th-232 and U-238 through descendants of the natural families: respectively Bi–212 (0.727 MeV) and Ti–208 (0.583 MeV) for thorium, Pb–214 (0.352 MeV) and Bi–214 (0.609–1.120 MeV) for uranium.

The data obtained were elaborated with an IBM personal computer equipped with the EG&G-Ortec programme "Quantitative analysis software programme for gamma spectrometry". Because of the low levels of Cs-137 contamination, measurements time was between 50 and 150 hours. Errors reported referred to standard deviation (1 sigma) and was due only to the counting statistics. All data were decay corrected to sampling time and are herein reported as Bq/m^3 for water samples and Bq/kg dry for solid samples.

The determination of the plutonium isotopes was carried out through a particular radiochemical separation[9] after a sample pre-treatment based on the matrix composition. Sediments were certainly the most analyzed samples and the procedure can be summarized in the following steps.

Ashed sediments were added of Pu–242 (4 dpm) as internal tracer and submitted to a HNO_3 attack. Different passages on microtene TOPO (oxyde of tri-n-octylphosphine) were executed and then a classical electrodeposition on steel plates in acid solution was carried out.

Electrodeposited plates were hence counted for alpha-spectrometry by means of two EG&G-Ortec silicon chambers and one ionizing chamber for the quali-quantitative identification of Pu-238 (5.5 MeV), Pu-239 + 240 (5.1 MeV) and Pu–242 (4.9 MeV).

The data obtained were elaborated with an IBM personal computer equipped with the EG&G-Ortec programme "Quantitative analysis software programme for alpha spectrometry" following the same operative procedures reported above for gamma counting.

Strontium was determined from the liquid residue of one passage of the plutonium extraction. In this technique yttrium (Sr–90 descendant) was extracted with HDEHP (di-(2-ethylhexyl)-phosphate) and precipitated with oxalic acid. Y–90 was therefore measured by means of an ASPN beta counter.

The organic matter contained in the sediment samples was determined as difference between ash (550°C) and dry weight; nitrogen with the Kjeldahl method; phosphorus through spectrophotometry and organic carbon with the COD method.

RESULTS AND DISCUSSION

The data detected allowed a complete and satisfactory mapping of the distribution of natural and artificial radioactivity in different ecosystem components of the Terra Nova Bay inlet and the forelying sea. In particular, higher concentrations of artificial radioactivity were measured in lacustrine algae samples (Tarn Flat, AG 12) and in mosses and lichens collected at Kay Island (VO 14). These samples were collected during the 1990–91 Campaign and the results are reported in Table 2.

Both aquatic and terrestrial vegetable organisms demonstrated to be efficient bioindicators of environmental radiocontamination of the 1990–91 study area. Stream waters derived from ice melting, besides their collection and stagnation in the small lake-pond system seemed to play a fundamental role in the transport and accumulation of Cs-137 and Pu-239,240 in lake sediments and algae. Their concentrations were generally higher than the ones detected in soil samples collected in areas surrounding the same lakes during the previous Campaigns[1–3,10].

Cs-137 and natural radionuclides concentrations measured in two soil cores (SL 8, Edmonson Point; SL 20, Skua Lake) and two lake sediment cores (LS 20 A e B, respectively center and side of Skua Lake) collected during the 1991–92 Expedition are illustrated in Table 3. The Cs-137 contribution to the core contamination was higher in the top layer (0–5 cm) whereas decreasing values were observed in the underlying strata (5–10 and 10–15 cm). In spite of this, level zero of radiocesium contamination was not detected due to the core shortness. For this reason inventory data were not calculated.

As far as natural radioactivity is concerned K-40 was the most abundant primordial radionuclide in all samples analyzed and concentrations of Th-232 and U-238 were in the ranges of previous data[1–3] in relation to the natural composition of soil and sediments.

Table 2 Concentrations of artificial and natural radioisotopes in environmental matrixes of the 1990–91 Campaign. Values in Bq/kg dry (data quoted with (*) are in Bq/mc).

Sample	*Cs-137*	*Sr-90*	*Pu-238*	*Pu-239,240*	*K-40*	*Th-232*	*U-238*
Lacustrine ecosystem							
LW 17/3 (*)	1.01 ± 0.65	< 0.001	< 0.001	< 0.001	ND	ND	ND
AG 8	5.97 ± 0.36	ND	ND	ND	406.6 ± 9.71	ND	ND
AG 12	46.8 ± 0.54	ND	ND	ND	568.5 ± 10.8	ND	ND
AG 17	10.4 ± 0.41	19.3 ± 1.5	0.03 ± 0.01	0.12 ± 0.01	204.3 ± 10.6	ND	ND
LS 8	3.21 ± 0.16	5.90 ± 0.4	< 0.02	0.08 ± 0.01	1130 ± 10.02	46.3 ± 1.5	53.3 ± 0.9
LS 12	0.14 ± 0.09	12.9 ± 1.1	< 0.02	0.02 ± 0.01	912.2 ± 7.91	58.1 ± 0.9	79.5 ± 1.2
LS 17	0.27 ± 0.19	0.3 ± 0.2	< 0.04	< 0.03	1014 ± 12.2	25.3 ± 1.2	56.9 ± 0.6
Terrestrial ecosystem							
SN 17 (*)	2.50 ± 1.12	0.01 ± 0.001	< 0.0001	< 0.0001	ND	ND	ND
VO 14 Lich.	47.5 ± 2.15	12.2 ± 1.0	0.05 ± 0.01	0.22 ± 0.03	239.8 ± 47.5	ND	ND
VO 14 Mos.	39.0 ± 0.64	13.3 ± 0.9	0.08 ± 0.01	0.30 ± 0.04	383.6 ± 13.3	ND	ND
SL 14 Mos.	1.49 ± 0.17	2.90 ± 0.2	< 0.04	< 0.03	771.8 ± 07.9	35.1 ± 1.3	42.3 ± 1.5

Table 3 Concentrations of Cs-137, K-40, Th-232 and U-238 in soil and lake sediment cores collected during the 1991–92 campaign. Values in Bq/kg dry.

Sample	*Stratum (cm)*	*Cs-137*	*K-40*	*Th-232*	*U-238*
Terrestrial ecosystem					
SL 8	0–5	0.71 ± 0.1	1334 ± 10.1	49.5 ± 1.0	64.2 ± 1.6
SL 8	5–10	0.22 ± 0.1	1066 ± 8.50	56.4 ± 0.9	46.1 ± 0.9
SL 8	10–15	0.15 ± 0.1	1153 ± 8.70	43.2 ± 0.8	53.5 ± 1.1
SL 20	0–5	0.79 ± 0.1	942 ± 06.9	38.3 ± 0.7	23.5 ± 0.4
SL 20	5–10	0.63 ± 0.1	1087 ± 7.66	41.2 ± 0.7	30.6 ± 0.5
SL 20	10–15	0.27 ± 0.1	945.9 ± 8.01	30.7 ± 0.6	23.7 ± 0.6
Lacustrine ecosystem					
LS 20A	0–5	0.23 ± 0.1	1042 ± 7.30	32.6 ± 0.6	24.5 ± 0.4
LS 20A	5–10	0.18 ± 0.1	985.6 ± 13.6	25.6 ± 0.7	20.5 ± 0.3
LS 20A	10–15	0.15 ± 0.1	1132 ± 10.9	25.9 ± 0.7	25.0 ± 0.3
LS 20B	0–5	0.26 ± 0.1	1189 ± 15.3	21.2 ± 0.7	25.7 ± 1.1
LS 20B	5–10	0.14 ± 0.1	1089 ± 16.2	23.9 ± 1.3	24.3 ± 0.9
Ls 20B	10–15	0.12 ± 0.1	1150 ± 15.9	25.1 ± 0.8	23.5 ± 0.3

Cs-137 concentrations detected in marine samples of the 1990–91 Campaign were generally lower than the ones observed in the lacustrine and terrestrial ecosystems. All analyses performed in the marine ecosystem were detailly described in previous papers[8,11]. In particular, ranges of concentrations of this radionuclide were 0.20–0.42 Bq/m^3 for sea water, 0.11–0.20 Bq/kg dry for coastal and off-shore sediments, 0.16–0.77 Bq/kg dry for the organisms different anatomical components.

Concentration factors (C.F.) for Cs-137 measured in marine organisms collected during the 1988–89, 1989–90 and 1990–91 Campaigns were generally high in the bivalve *Adamussium colbecki* in relation to the filter—feeding habit of this species. On the contrary, the ones determined in *Pagothenia bernacchii* and *Chionodraco hamatus* were lower and reflected different feeding habits than *Adamussium colbecki*. Ranges of the C.F. determined in total organisms and single components are reported in Figure 1. Nevertheless, the concentration factors determined in *Adamussium* seemed quite well related to the seasonality of the ecosystem with higher values during the austral summer periods of intense primary production.

This might suggest a major role of the particle form of Cs-137 in transporting this isotope to the Antarctic benthic compartment, although it is well known that radiocesium in temperate waters is mostly in the soluble form.

The trends of Cs-137 and organic matter in marine coastal and off shore sediments seemed to support this idea. In fact, higher Cs-137 concentrations were detected in samples containing higher values of organic matter (5.5–8.0% dry weight; average value other samples 3.86%), organic carbon (3.61–7.88 mg/g; average value other samples 1.86 mg/g) and nitrogen (3.01–4.60 mg/g; average value other samples 1.11 mg/g). Concentrations of Cs-137 were quite well correlated with the contents of the above elements. In fact, their correlation coefficient ranged between 0.65 and 0.95.

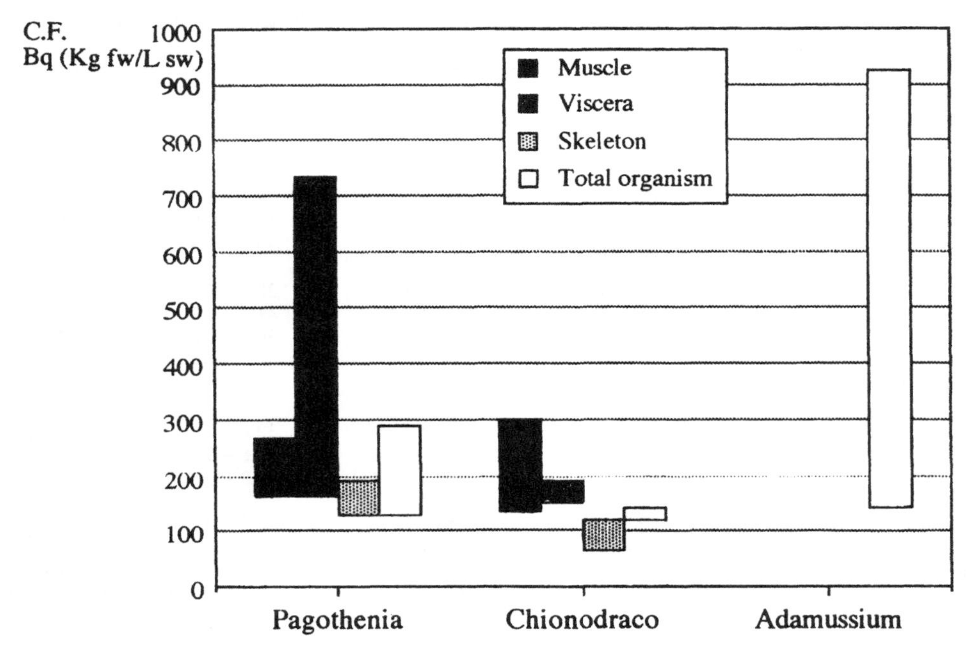

Figure 1 Ranges of concentration factors determined in single components and total organisms of Antarctic marine fauna.

References

1. C. Triulzi, A. Mangia, A. Casoli, S. Albertazzi and F. Nonnis Marzano, *Annali di Chimica*, **79**, 723–733. (1989).
2. C. Triulzi, A. Mangia, A. Mori and F. Nonnis Marzano, *Proceedings Meeting Environmental Impact in Antarctica*, (Rome, June 8–9, pp 133–141. 1990).
3. C. Triulzi, F. Nonnis Marzano, A. Mori, A. Casoli and M. Vaghi, *Annali di Chimica*, **81**, 549–561. (1991).
4. C. Triulzi, F. Nonnis Marzano and A. Mori, *Proceedings 9th Meeting AIOL*, (Genova, pp. 583–592. 1992).
5. W. Martinotti, F. Nonnis Marzano, G. Queirazza and C. Triulzi, *Proceedings International Conference on Environmental Radioactivity in the Arctic and Antarctic*, (Kirkenes, Norway, August 23–27, pp. 233–236. 1993).
6. H. G. Petrow and H. Levine, *Anal. Chem.*, **39**, 360–362. (1967).
7. Relazioni finali inerenti le Spedizioni 1990–91 e 1991–92. Settore Impatto Ambientale e Metodologie Chimiche. Università di Venezia, Bologna e Genova.
8. F. Nonnis Marzano and C. Triulzi. *Intern. J. Environ. Anal. Chem.*, **55**, 243–252. (1994).
9. A. Delle Site, V. Marchionni, C. Testa and C. Triulzi, *Anal. Chim. Acta.*, **117**, 217–224. (1980).
10. C. Triulzi, F. Nonnis Marzano, A. Mori, A. Casoli and M. Vaghi, *Proceedings 2nd Meeting. Environmental Impact Chemical Methodologies*, (Venice, May 26–28, pp. 112–116. 1992).
11. O. Tubertini, M. G. Bettoli, L. Cantelli, L. Tositti, S. Valcher, C. Triulzi, F. Nonnis Marzano, A. Mori, M. Vaghi, G. Sbrignadello, S. Degetto and M. Faggin, *Proceedings International Conference on environmental radioactivity in the Arctic and Antarctic*, (Kirkenes, Norway, August 21–28, pp. 195–199. 1993).

CHARACTERIZATION OF HUMIC ACIDS ISOLATED FROM ANTARCTIC SOILS

N. CALACE, L. CAMPANELLA, F. DE PAOLIS and B. M. PETRONIO*

Department of Chemistry, University of Rome "La Sapienza" piazzale Aldo Moro 5, 00185 Rome, Italy

Humic acids isolated from Antarctic soils have been characterized by FTIR, ^{13}C-NMR and elemental analysis; aminoacid content has been determined after hydrolysis with 6 N HCl. Nitrogen and hydrogen contents of HAs are very high, indicating that more proteinaceous material, covalently bound to the structures, has not undergone remarkable transformation during the humification processes. Moreover, the aliphatic carbon prevails on the aromatic one, owing to the nature of non-ligniferous material as precursor. With the progress of the humification process, obtained in temperate climatic conditions, the proteinaceous material decreases, as shown by the analysis of aminoacids, and the molecular weight of the humic compounds increases whereas aromatic carbon does not.

KEY WORDS: Antarctic soils, humic acids.

INTRODUCTION

The main characteristic of humic substances is the presence of high molecular weight polimeric structures.

Schnitzer *et al.*[1] have recently suggested a carbon skeleton in which alkylbenzene structures play a dominant role in soil humic acids (HA). Because the humification processes strictly depend both on the parent materials and on the physico-chemical and biological parameters, the structures of soil humic acids from Antarctica are probably different from those ones suggested by Schnitzer.

The absence of lignin as precursor may be responsible for a higher proportion of aliphatic structures. Moreover, the lack of humidity may determine a greatest amount of large molecules, because the depolymerization processes are slower. Humic molecules, apparently polymerize progressively during the dry season, while the duration of the humid season appears to be responsible for the decrease in the amount of large molecules because of the depolymerization process[2].

In this paper we have studied some humic acids coming from Antarctic soils, their structures characterized by FTIR, ^{13}C-NMR, elemental analysis have been compared with those ones reported in literature.

* Corresponding author

EXPERIMENTAL

Sampling

The soil samples have been collected in the zones listed in Table 1. The Nos. 7 and 8 samples come from damp microclimate zones (near flowing waters and lakes where moss grows profusely).

All samples are superficial (0–15 cm) because under the 0–15 cm layer the permafrost is present; only in the stations Nos. 1 and 9 are collected both superficial (0–15 cm) and lower (15–30 cm) layers.

Procedure

The soil samples were frozen immediately after the collection, then defrozen (six months after), sieved (4 mm) and treated according to the procedure of International Humic Substances Society for extraction of humics, as outlined previously[3].

In the case of sample No. 7, a soil portion was defrozen and then held in temperate climatic condition (30–38°C, regularly damped with water) for three months; humic and fulvic acids were then extracted.

Humic and fulvic acids were characterized by FTIR and ^{13}C-NMR spectrophotometry and elemental analysis; the concentration of carboxyl groups was determined with the calcium acetate method[4].

In some cases both the determination of aminoacids present as component in the humic structures (carried out after hydrolysis with 6N HCl[5]) and the distribution of molecular weights by gel permeation chromatography on Sephadex G resins[6] have been carried out.

Apparatus

A FTIR Philips spectrophotometer model P3202 working in diffuse reflectance conditions was used. The results are given in Kubelka Munk units; the Kubelka Munk is

Table 1 Location of sampling points and extraction yields (%).

Sample No.	*Lat S*	*Long E*	*Yields (%)*	
			HA	*FA*
1	70°20'	165°07'	0.003	0.07
2	70°20'	165°07'	0.08	0.09
3	74°47'	163°38'	0.02	0.07
4	74°43'	164°01'	0	0.06
5	74°43'	164°01'	0.06	0.07
6	74°59'	162°33'	0	0.05
7	74°59'	162°33'	0.13	0.28
8	74°41'	164°02'	0.24	0.17
9	74°41'	164°02'	0.009	0.05

a mathematical formula applied to diffuse reflectance spectra. The samples were prepared by mixing the dried humic acids (1 mg) with anidrous KBr (100 mg).

^{13}C-NMR spectra were determined using a Varian spectrometer model XL-300. The samples were prepared in a NMR tube (5 mm) by dissolving the dried humics (30 mg) in 1 ml of NaOD 0.5 M. The operating conditions were: 75 MHz, pulse 45°, acquisition time 0.1 sec, delay time 0.5 sec. From 600.000 to 800.000 scans were accumulated.

Elemental analyses were carried out by a Carlo Erba 240-B model CHN-analyzer in the Microanalysis Laboratory of the Italian Research Council.

A Pharmacia LKB plus 4151 aminoacid analyzer was used, equipped with a cationic exchange resin. Elution was carried out with buffer solution (citric acid/citrate) at different pH values. Aminoacids were detected with ninhydrin reagent.

RESULTS

Both humic and fulvic acids were extracted from samples collected in damp microclimate zones (samples n. 7 and 8); in the other zones, sometimes only fulvic acids were present. However, in all the samples (except Nos. 7 and 8) the humified matter is poor. Comparing HA extracted from superficial samples (0–15 cm) and those from the corresponding layer underneath (15–30 cm) some differences are evident. The extraction yield for the superficial sample is lower (0.003% and 0.02% for sample n.1 and 0.009% and 0.03% for sample n.9), as observed by several authors in soils with no illuvial horizon[7,8]. Moreover, FTIR and ^{13}C-NMR spectra show that the superficial HA contain a greater concentration of carboxyl groups (172 ppm) and a higher content of nitrogenous compounds (1660–1540 cm^{-1}), confirmed by N/C ratio (Table 2), probably reflecting aminoacids bound to the humic structures. Aliphatic carbon is present especially as long chains of fatty acids (peak at 32 ppm). On the contrary, HA from the deeper layer have a more heterogeneous composition. In the aliphatic zone (0–50 ppm) peaks corresponding to terminal methyl groups (0–20 ppm) are shown, and in the 20–50 ppm range the signals have equal intensity (Figure 1). Carbon bonded to O or N heteroatoms is evident.

The differences observed may be ascribed to the different stages of the humification processes that in the deeper layer are more advanced. In fact, it is known that the humification processes are accompanied by some chemical modifications such as loss of –COOH functional groups, dehydrogenation, ring closure, increase of aromaticity[9]. The

Table 2 Elemental analysis of humic acids extracted from different layers of soils: superficial (0–15 cm) (A) and (15–30 cm) (B).

Sample n.	*1*		*9*	
	A	*B*	*A*	*B*
N (%)	8.57 + 0.34	3.96 + 0.16	8.04 + 0.21	3.38 + 0.08
C (%)	53.68 + 2.15	38.70 + 1.55	60.29 + 2.41	38.07 + 1.52
H (%)	8.24 + 0.41	6.93 + 0.41	8.04 + 0.32	4.02 + 0.20
O (%)	29.51 + 1.47	50.41 + 2.21	23.63 + 0.62	54.23 + 2.01
H/C	1.84 + 0.07	2.15 + 0.09	1.60 + 0.05	2.68 + 0.11
N/C	0.14 + 0.01	0.09 + 0.01	0.11 + 0.02	0.08 + 0.11
Ash (%)	12.3 + 2.0	14.0 + 3.0	25.2 + 3.0	23.5 + 2.5

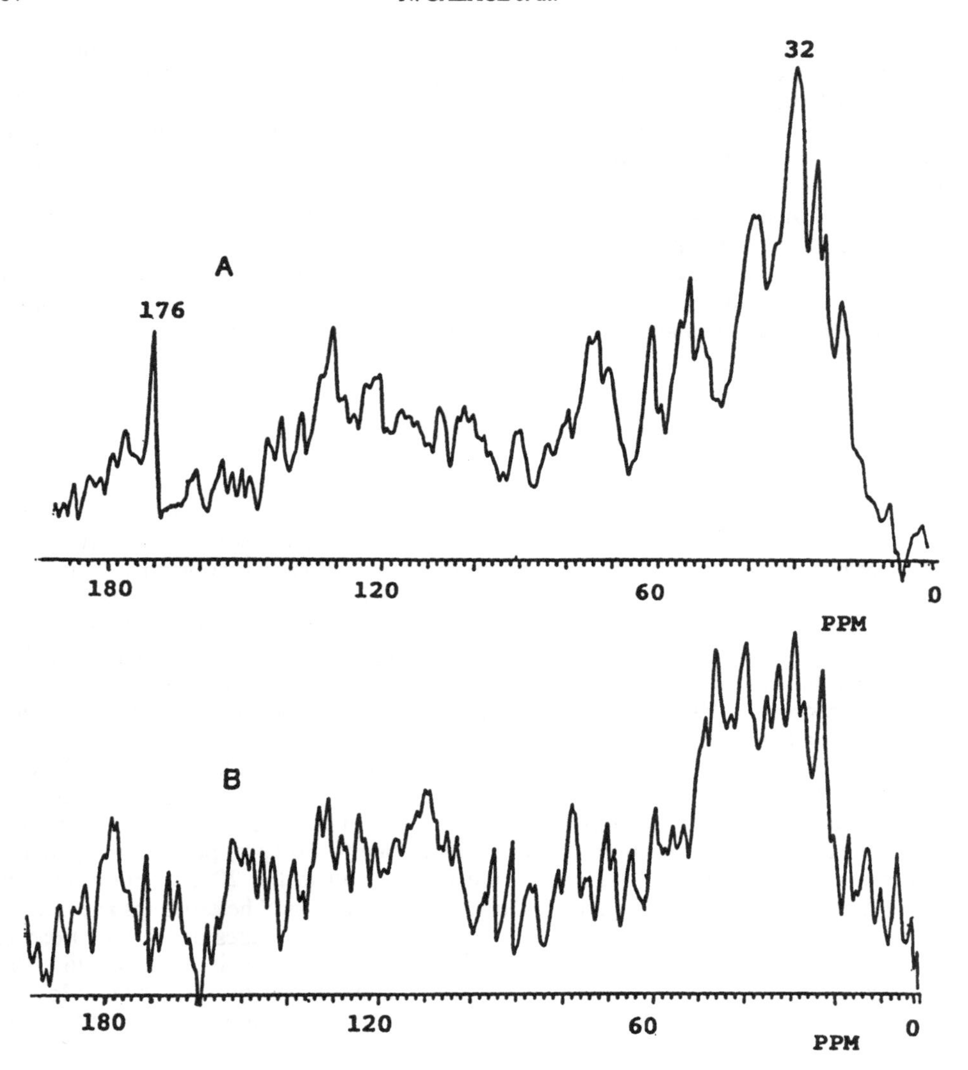

Figure 1 ^{13}C-MNR of HA of superficial (A) and upper (B) layer.

elemental analysis data for humic acids of the lower horizon of soil show that the percentage of C, H, and especially N decrease (Table 2).

Analysing the FTIR and ^{13}C-NMR spectra of humic acids obtained from the superficial samples, some differences are observed between those having mosses (samples Nos. 7 and 8) as precursors and the others. On the contrary, the data of elemental analysis (Table 3) do not differ significantly: only the oxygen content is higher (samples 7 and 8).

The amount of humified matter, both HA and FA, is higher in samples Nos. 7 and 8 (derived by moss), (Table 1) and it is even smaller than that obtained from no Antarctic soils[10].

Table 3 Elemental analyses of humic acids extracted from superficial (0–15 cm) layer of soil samples.

Sample No.	*2*	*3*	*5*	*7*	*8*
N (%)	10.41 + 0.91	8.48 + 0.52	11.30 + 0.67	7.52 + 0.30	8.86 + 0.53
C (%)	66.80 + 2.67	63.20 + 2.14	68.12 + 2.58	59.62 + 1.49	56.02 + 1.68
H (%)	8.93 + 0.27	7.62 + 0.24	9.10 + 0.23	8.17 + 0.16	7.65 + 0.19
O (%)	13.90 + 0.82	20.72 + 0.62	11.53 + 0.57	24.70 + 1.02	27.52 + 1.03
H/C	1.61 + 0.10	1.44 + 0.11	1.60 + 0.09	1.64 + 0.11	1.64 + 0.11
N/C	0.13 + 0.03	0.11 + 0.02	0.14 + 0.02	0.11 + 0.01	0.13 + 0.03
COOH (meq/g)	1.34 + 0.02	1.78 + 0.13	1.78 + 0.14	1.36 + 0.02	1.86 + 0.07
Ash (%)	17.7 + 3.5	22.2 + 3.0	21.4 + 2.5	18.5 + 4.0	10.1 + 2.5

FTIR and ^{13}C-NMR spectra of HA of samples Nos. 7 and 8 show a high adsorption at 1715 cm^{-1} and a signal at 190 ppm, probably due to C = 0 groups of aldehydes and ketones, since the concentration of carboxylic groups is not very high.

However, all Antarctic superficial HA are characterized by the aliphatic carbons prevailing over the aromatic ones. This characteristic can be related with the nature of precursors. In Antarctica the humic substances rise from mosses and algae materials rich in lipids, carbohydrates and proteins, without lignin. The presence of aromatic structures could be prevalently due to residual aromatic aminoacids like tryptophan, phenylalanine and tyrosine of algal origin or to tannins and flavonoids from different origin, or to both of them.

Nitrogen percentage of Antarctic HA is higher (Table 3) than that of the corresponding HA of no Antarctic soils[11–16], probably due to slow transfromations of protein and peptides during the humification processes. The oxygen content and carboxyl groups concentration are generally lower, because of the nature of precursors and the particular climatic conditions. Consequently, in Antarctica the structures of HA of soils are similar to those of marine and lacustrine sediments[3,17–19].

To study the effects of climatic conditions on the Antarctic humic substances we have compared humic acids extracted from sample No. 7 (A) with those of the same sample mantained three months at temperate conditions (30–38°C) (B).

The extraction yield increases (0.13% for A and 0.29% for B), the elemental analysis data do not change significantly and the major oxygen containing groups decrease. The total content of aminoacids, obtained after hydrolysis with 6 N HCl, decreases (2.55 and 2.07 μmol/mg respectively; the signals at 1660 and 1540 cm^{-1} of FTIR spectra (Figure 3) are inverted). Aspartic and glutamic acids decrease significantly (Figure 2). In both samples the percentages of acid, basic and neutral aminoacids are in good agreement with those reported by Stevenson[20] for many soils.

The distribution of molecular weights and the nature of the single fractions are different for the two samples (Table 4). In samples the compounds of the first fraction (1.500–5.000 molecular weight range) show a little content of nitrogen (N/C rate 0.01 instead of 0.08) and a higher percentage of C = C stretching vibrations (H/C rate 0.71 instead of 1.33). ^{13}C-NMR spectrum is not well resolved indicating the complexity and the no repeatibility of the structure of the sample subunits[21,22]. FTIR spectrum shows a signal at 1760 cm^{-1}, probably due to C = 0 stretching of β-ketoesters; in the corresponding fraction of sample A (Figure 3) the prevailing signal appears at 1710 cm^{-1} (carboxyl groups) according to the elemental analysis data. Compounds with a high molecular weight (> 100.000) are present only in sample B (Table 4).

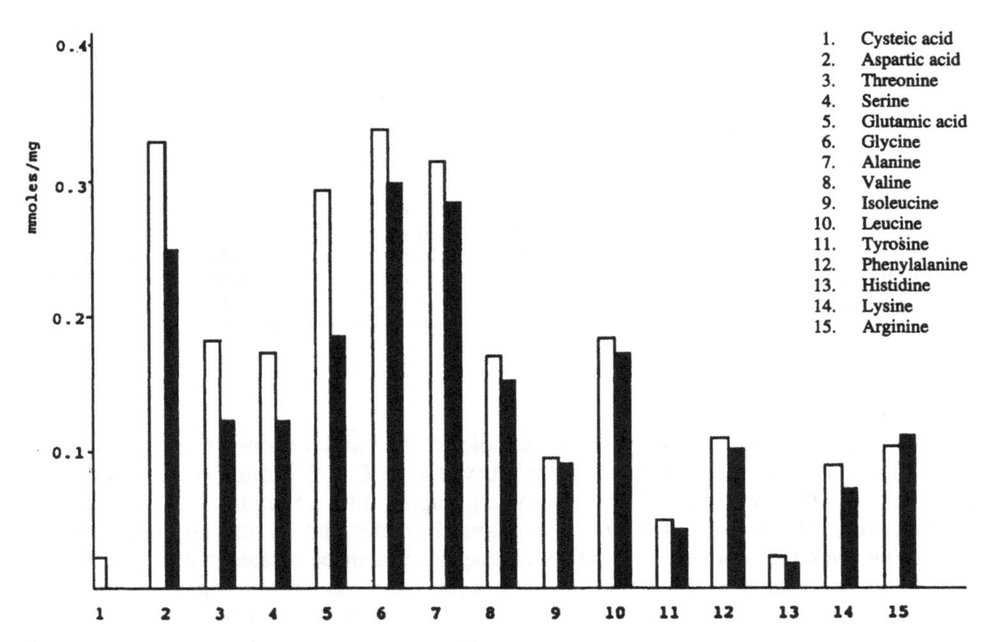

Figure 2 Aminoacid distribution in HA of sample n. 7 before □ and after ■ the progress of humification process.

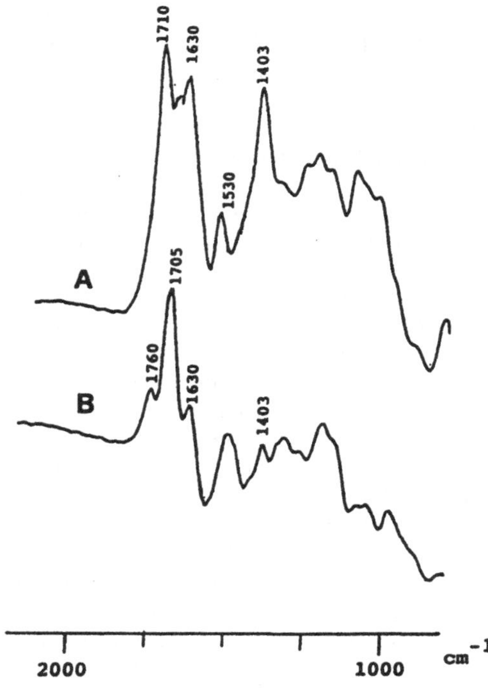

Figure 3 FTIR spectra of HA of the fractions with molecular weight 1500–5000 before (A) and after (B) the progress of the humification process.

Table 4 Elemental analysis, major containing oxygen groups and yields of humic acids extracted from sample No. 7 before (A) and after three months in climatic temperate conditions (B).

Sample No.	*7A*	*7B*
Yield (%)	0.10 + 0.05	0.30 + 0.05
H (%)	8.17 + 0.16	8.40 + 0.24
N (%)	7.52 + 0.30	7.05 + 0.25
C (%)	59.62 + 1.49	61.45 + 2.50
O (%)	24.69 + 1.24	23.10 + 2.05
H/C	1.64 + 0.11	1.64 + 0.11
N/C	0.11 + 0.01	0.10 + 0.01
O/C	0.31 + 0.01	0.28 + 0.01
Ash (%)	18.5 + 2.5	22.0 + 3.5
Aminoacid (nmol/mg)	2552 + 80	2072 + 75
Total acidity (meq/g)	6.06 + 1.03	3.73 + 0.76
–OH groups (meq/g)	4.70 + 1.04	2.71 + 0.77
–COOH groups (meq/g)	1.36 + 0.01	1.02 + 0.01

Table 5 Differences between HA (samples No. 7A and 7B) with different degree of humification (A and B).

		7A			*7B*	
Molecular weight	*Yield*	*N/C*	*H/C*	*Yield*	*N/C*	*H/C*
1500–5000	5.6	0.08	1.33	5.8	0.01	0.71
50000–100000	90.2	0.13	1.50	9.0	0.12	1.47
> 100000	–	–	–	84.0	0.11	1.58

The differences observed point out that the humified process proceeds with the formation of compounds characterized by molecules with high molecular weight. In this process the most abundant fraction (molecular weight 50000–100000) of sample A is probably involved. Moreover, the increase of extraction yield suggests that non-humic materials begin the humified process; the formation of compounds with low molecular weight may be due to the short exposition time.

CONCLUSIONS

Antarctic soils generally contain a low amount of humic compounds, owing to the lack of organic materials.

Humic compounds are present in high amounts only near the lakes and flowing waters there is a high amount of parent materials; in the other sites humic material is present in low quantity because it comes from the snow precipitations[23] or it is derived from the organic matter transported by the wind.

From an environmental point of view, the results obtained point out that the complexing capacity of Antarctic soils towards metals, due to the organic matter, is low. Moreover, the complexing capacity of Antarctic HA could be different owing to the differences observed in the structures.

Acknowledgments

The present work was realized with finantial support of E.N.E.A. (Rome) under the Italian Antarctic Project, section 2d.3: Environmental Contamination.

The authors express their appreciation to Dr. Eugenia Schininà (Department Biochemical Science "A. Rossi Fanelli", University "La Sapienza" Rome) for aminoacids analysis and to Franco Dianetti and Lucantonio Petrilli (Italian National Council of Research Monte Libretti, Rome) for elemental analysis.

References

1. H. R. Schulten and M. Schnitzer: *Naturwissenshaften*, **80**, 23–30 (1993).
2. K. Nakane, *Jpn. J. Ecol.*, **30**, 19–29 (1980).
3. L. Campanella, T. Ferri, B. M. Petronio, A. Pupella and M. Paternoster, *Ann. Chim.*, **81**, 477–490 (1991).
4. M. Schitzer and V. C. Gupta, *Soil Sci. Soc. Amer. Proc.*, **29**, 274–277 (1965).
5. F. J. Stevenson and K. M. Goh, *Geochim. Cosmochim. Acta*, **35**, 471–483 (1971).
6. R. L. Wershaw, D. J. Pinckney, E. C. Llaguno and V. Vicente-Beckett, *Anal. Chim. Acta*, **232**, 31–42 (1990).
7. V. V. Ponomareva, *Sov. Soil Sci.*, **6**, 393–402 (1974).
8. D. Schwartz, B. Guillet, G. Villemin and F. Toutain, *Pédologie*, **36**, 179–198 (1986).
9. A. Nissembaum and I. R. Kaplan, *Limnol. Oceanogr.*, **17**, 570–582 (1972).
10. G. Calderoni and M. Schnitzer, *Geochim. Cosmochim. Acta*, **248**, 2045–2051 (1984).
11. F. J. Stevenson and K. M. Goh, *Soil Sci.*, **113**, 334–345 (1972).
12. V. O. Biederbeck and E. A. Paul, *Soil Sci.*, **115**, 357–366 (1973).
13. P. G. Hatcher, M. Schnitzer, L. W. Dennis and G. E. Maciel, *Soil Sci. Soc. Am. J.*, **45**, 1089–1094 (1981).
14. R. Frund, H. D. Ludemann, F. J. Gonzales-Villa and G. Almendros. *Sci. Total Environ*, **81/82**, 187–194 (1989).
15. N. Senesi, T. M. Miano and M. R. Provenzano, *Sci. Total Environ.*, 143–156 (1989).
16. R. A. Rossel, A. E. Andriuolo and M. Schnitzer, *Sci. Total Envrion.*, 391–400 (1989).
17. L. Campanella, T. Ferri, B. M. Petronio, A. Pupella, M. Soldani and B. Cosma, *Ann. Chim.*, **81**, 417–436 (1991).
18. L. Campanella, T. Ferri, B. M. Petronio, A. Pupella and F. Tedesco, *Ann. Chim.*, **81**, 511–521 (1991).
19. L. Campanella, B. Cosma, N. Degli Innocenti, T. Ferri, B. M. Petronio and A. Pupella, *Intern. J. Environ. Anal Chem.*, **55**, 61–75 (1994).
20. F. J. Stevenson, *Humus chemistry, Genesis, Composition, Reactions* (Wiley, New York, 1982).
21. D. H. Stuermer and J. R. Payne, *Geochim. Cosmochim. Acta*, **40**, 1109–1111 (1976).
22. P. Ruggiero, O. Sciacovelli, C. Testini and F. S. Interesse, *Geochim. Cosmochim. Acta*, **42**, 411–416 (1978).
23. R. Cini, B. M. Petronio, N. Degli Innocenti, A. M. Stortini,C. Braguglia and N. Calace, *Ann. Chim.*, **84** (1994) in press.

SEDIMENTARY HUMIC ACIDS IN THE CONTINENTAL MARGIN OF THE ROSS SEA (ANTARCTICA)

C. M. BRAGUGLIA, L. CAMPANELLA, B. M. PETRONIO* and R. SCERBO

Department of Chemistry, University of Rome "La Sapienza" piazzale A. Moro 5, 00185 Rome, Italy

Humic acids (HA) extracted from Antarctic marine sediments were characterized using different techniques (FTIR, NMR, Thermogravimetry, Elemental Analysis and Ionic Exchange Chromatography). Antarctic sedimentary HA show a greater content of nitrogen and a predominant aliphatic nature when compared with data reported in the literature for non Antarctic HA samples. Some differences in the structures, such as the presence of aromatic carbon rings, carbohydrates, fatty acids, etc., have been observed, as a function of sampling zones. The differences are probably due to a different humification process, possibly depending on different trends of temperature and salinity measured down to the water column.

KEY WORDS: Sedimentary humic acids, Antarctica, Ross Sea

INTRODUCTION

The organic part of the sediments may be divided in humic (humic and fulvic acids) and non-humic fractions (carbohydrates, lipids, aminoacids, organic acids, etc.).

The humic fraction is the prevailing part[1] and consists of an heterogeneous mixture of compounds, yellow to black colored, having high molecular weight. Aminoacids and carbohydrates are also part of the humic structures. Oxygen is present within different functional groups (e.g., COOH, OH, C = O, and ethers and esters).

The humic acids extracted from marine sediments may have autochtonous and allochtonous origin, according to some authors[2]. As regards the autochtonous formation a condensation (probably like Maillard's) of carbohydrates, aminoacids and small molecules, has been hypothized[3]. Beside this hypothesis, Harvey *et al.*[4] suggest a mechanism in which fatty acids or polyunsatured glycerides, through an oxidative cross-linking, give rise to fulvic acids first and then humic acids.

The characterization of the structures of humic compounds provides important information about the primary material and the probable formation's mechanism. It represents the starting point to evaluate the capacity of interaction towards metals and different classes of organic compounds (e.g. herbicides and pesticides). In fact, the great

* Corresponding author.

reactivity of humic substances must be connected with the high number of functional groups: their presence depends both on the humification process and the organic precursors.

The study of humic compounds in Antarctic sediments is important because the starting material is non-ligniferous and the formation's mechanism may be feasibly affected by the typical climatic conditions of Antarctica (very low temperatures, alternation of long periods of light and darkness, lack of humidity, etc...).

EXPERIMENTAL

Sampling

Sediment samples were collected with box-corer in the Antarctic expedition in 1990/91; for each station a superficial (0–15 cm) and the layer below (15–30 cm) of the sediment have been sampled. Immediately after collection the samples were frozen at –30°C.

The sediment samples collected were divided into three groups according to the temperature trend of the water column[5]. Humic acids (HA) from sediments of two groups: n.1 and 2 (zone A) and n.3–6 (zone B) (Table 1) have been analyzed.

Procedure

Fulvic (FA) and humic (HA) acids were extracted from the sediments with the procedure outlined by the International Humic Substances Society[6] with the modifications suggested by Rashid and King[7]. After the NaOH treatment (solubilization of FA_I and HA_I), a series of alternate extractions with HCl 0.1 M and NaOH 0.5 M is carried out in order to solubilize the humic fraction bound to the clay through metal ions (FA_{II} and HA_{II}).

At first time, the I and II fractions were not mixed and were analyzed for elemental composition, thermogravimetry, FTIR and ^{13}C-NMR spectrometry.

Then the two fractions were mixed and again analyzed. Some (HA) samples were subjected to an acid hydrolysis (HC1 6 N) and aminoacids in the hydrolized solutions were identified and quantified by ionic exchange chromatography.

Table 1 Sample location and water depth.

Sample	*Lat.S*	*Long.E*	*Water depth*
1	70°43'4	171°28'3	2285 m
2	70°45'9	172°51'4	2410 m
3	74°25'68	169°35'28	678 m
4	75°01'8	169°56'7	336 m
5	74°38'6	171°32'2	463 m
6	74°46'5	168°10'05	495 m

Apparatus

A FTIR Philips spectrophotometer model P3202 working in diffuse reflectance conditions was used. The results are given in Kubelka Munk units; the Kubelka Munk is a mathematical formula applied to diffuse reflectance spectra. The samples were prepared by mixing the dried humic acids (1 mg) with anhydrous KBr (100 mg).

^{13}C-NMR spectra were determined using a Varian spectrometer model XL–300. The samples were prepared in a NMR tube (5 mm) by dissolving the dried humics (30–10 mg according to the available material) in 1 ml of NaOD 0.5 M. The operating conditions were: 75 MHz, pulse 45°, acquisition time 0.1 sec., delay time 0.5 sec. From 600,000 to 1000,000 scans were accumulated, according to the sample concentration.

Elemental analysis were carried out by a Carlo Erba 240–B model CHN-analyzer in the Microanalysis Laboratory of the Italian Research Council.

Thermogravimetric analysis were carried by a Perkin Elmer TGA thermogravimetric analyzer in N_2 atmosphere, between 50 and 950°C, scanning rate 20°C min^{-1}, 1.5 mg of sample.

A Pharmacia LKB plus 4151 aminoacid analyzer was used, equipped with a cationic exchange resin. Elution was carried out with buffer solutions (citric acid/citrate) at different pH values. Aminoacids were detected with ninhydrin reagent.

RESULTS AND DISCUSSION

Fulvic acids have not been obtained, as observed for Antarctic marine sediment samples[8,9].

Comparison of the ^{13}C-NMR and FTIR spectra (Figure 1) and elemental analysis data of HA_I and HA_{II} (N%, 7.8 and 7.2; C%, 72.2 and 67.8; H%, 8.1 and 7.9, respectively) shows that they are similar; consequently, we have applied the Rashid and King extraction procedure blending HA_I and HA_{II}.

Results reported in Table 2 are referred to the mixture of HA_I and HA_{II}. In this Table are also reported data ranges of the literature[10–13].

The extraction yields for the Antarctic sediments are lower than those of the non-Antarctic ones[2], even though the extraction method we adopted followed the usual scheme, with the addition of a further step. The elemental analyses show that the nitrogen content is higher than the values of the literature, while generally that of oxygen is lower; the H/C ratio is higher.

The FTIR spectra of the humic acids before and after hydrolysis demonstrate that the prevailing part of the nitrogen content is due to the aminoacids bound to the humic structure[14]. In fact, the bands at 1535–1540 cm^{-1} (due to the stretching of peptidic linkage) and at 1690 cm^{-1} (I band of amides) result drastically reduced, whereas the one at 1710 cm^{-1} (stretching C = O of carboxyl groups) is unchanged (Figure 2). Moreover, the N/C ratio decreases after hydrolisis (from 0.1 to 0.05). ^{13}C-NMR spectra of an Antarctic and a non-Antarctic humic acid[15] (Figure 3) show that in the last one the ratio between aliphatic carbon (0–50 ppm) and the aromatic one is lower. Moreover, the 170 and 200 ppm zone, typical of carboxyl, ketonic, amidic and aldehydic groups is higher (if compared with the aliphatic carbon zone), according to the high oxygen content of the non-Antarctic humic acids.

Humic compounds shortage in Antarctic sediments may be attributable both to the poorness of organic substrates and to the rigid climate of Antarctica that makes slower

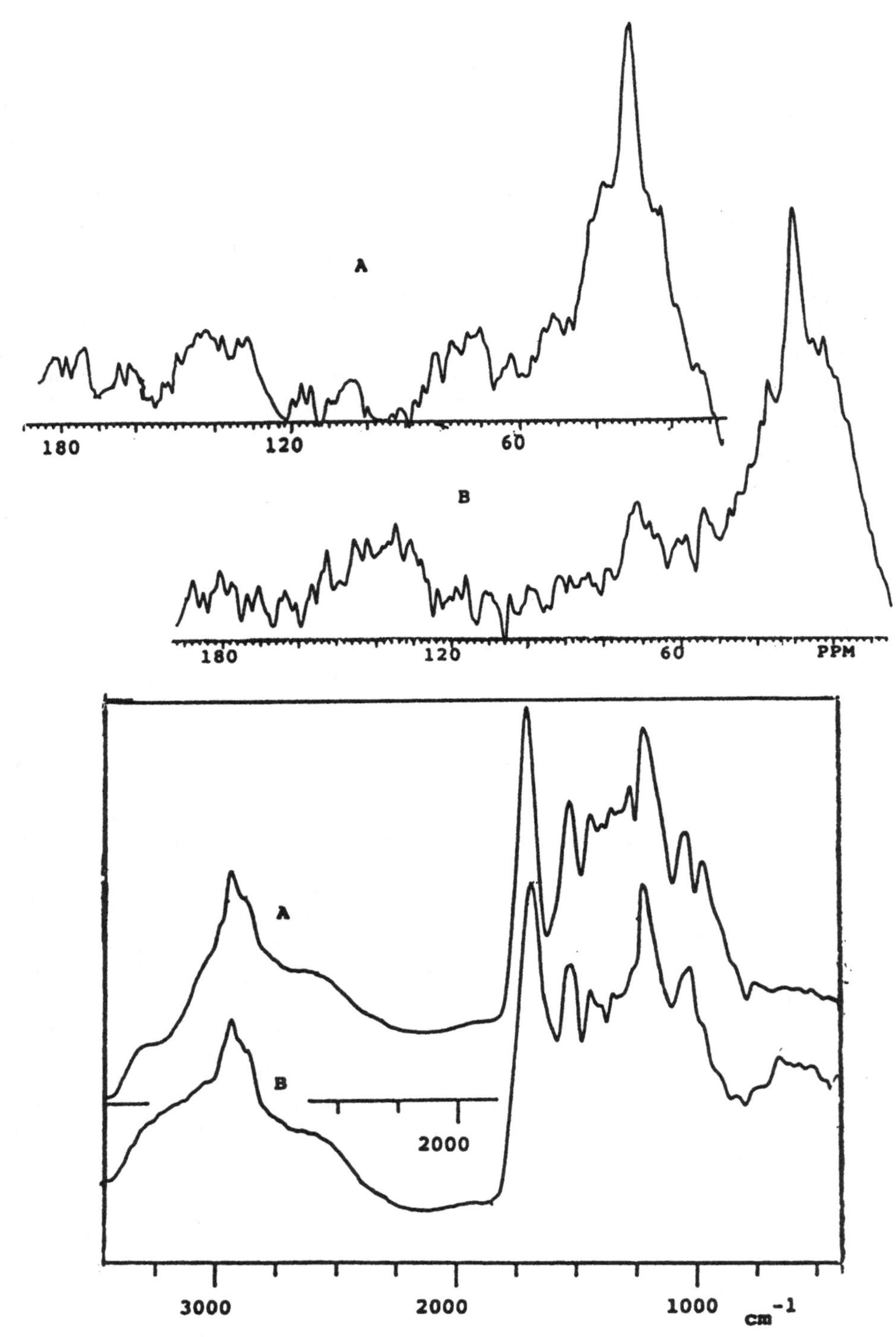

Figure 1 FT-IR and ^{13}C-NMR spectra of HA_I (A) and HA_{II} (B).

Table 2 Elemental analysis (%) and yields (%) of HA (ash- free) extracted from superficial and upper layers of sediment samples.

Sample	*Layer*	*N (%)*	*C (%)*	*H (%)*	*O (%)*	*H/C*	*N/C*	*Yield %*
1	superf.	5.98	51.97	6.05	36.0	1.40	0.099	0.028
	upper	5.81	51.16	5.37	37.7	1.26	0.097	0.005
2	superf.	6.78	56.78	7.14	29.3	1.51	0.102	0.043
	upper	6.48	62.23	6.39	24.9	1.23	0.090	0.020
3	superf.	6.15	79.75	6.85	7.25	1.03	0.070	0.082
	upper	5.46	77.46	6.72	10.40	1.04	0.060	0.056
4	superf.	8.23	72.71	9.00	10.1	1.49	0.100	0.187
	upper	6.49	77.99	7.78	7.74	1.20	0.070	0.059
5	superf.	7.10	73.71	8.57	10.6	1.40	0.090	0.095
	upper	7.02	72.56	8.40	12.0	1.39	0.080	0.373
6	superf.	7.53	67.05	8.02	17.4	1.44	0.100	0.231
	upper	5.88	51.81	6.41	35.9	1.48	0.100	0.402

Literature:	*N (%)*	*C (%)*	*H (%)*	*O (%)*	*H/C*	*Yield %*
superf.	3.9–5.5	53–60	4.8–7.1	26–38	1.1–1.5	0.4–0.8
upper	–	–	–	–	–	0.5–0.8

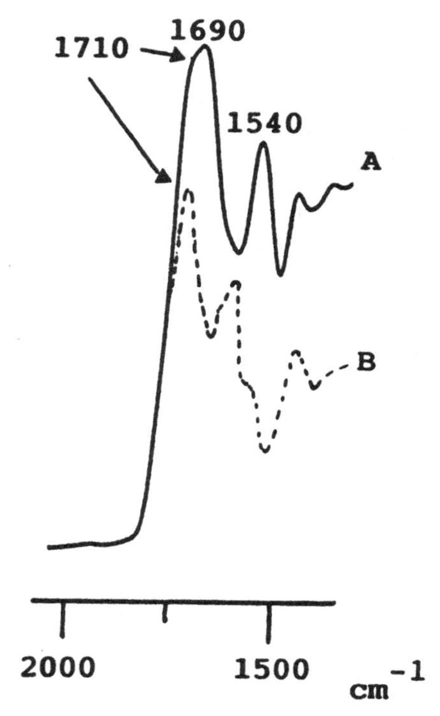

Figure 2 FT-IR spectra of humic acids before (A) and after (B) acid hydrolisis.

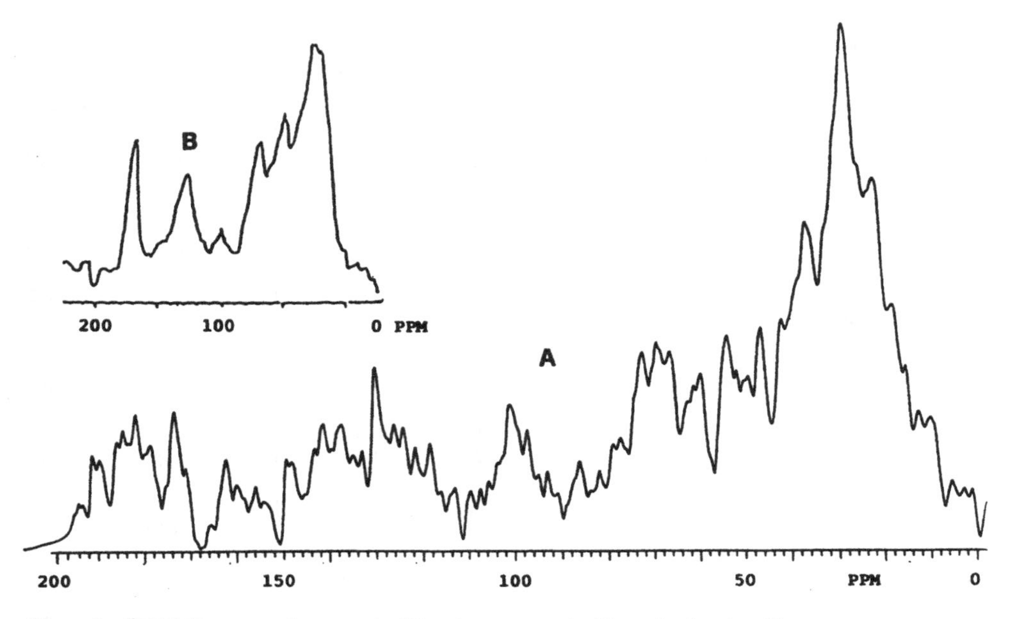

Figure 3 ^{13}C-NMR spectra of an antartic (A) and a non-antartic (B) marine humic acid.

the microbiological activity and the chemical transformations. Also the high nitrogen and the lower oxygen contents may be attributable to an incomplete transformation of proteinaceous materials due both to biological activity considerably reduced[16] and to slow chemical or enzymatic oxidation processes of aliphatic chains.

The low degree of condensation and aromatization, typical of the humic material in marine sediments[17], is particularly marked for the Antarctic marine humic acids (deducible from H/C ratio and ^{13}C-NMR spectra). In fact, the allochthonous contribution from the continent cannot influence the aromaticity of the humic substances, because the terrestrial Antarctic flora is made up to algae and moss devoid of lignin, aromatic precursor of the humic material.

Comparing the values of elemental analysis (Table 2) of humic acids extracted from the superficial and from the upper layer, it may be observed that the humic acids of the superficial layer have higher nitrogen and hydrogen contents. Humic acids of the superficial layer seems to show a lower sedimentary alteration's degree; this process occurs with loss of aliphatic chains and of nitrogen as ammonia, for mineralization of proteins[18,19]. A decrease of functional groups (Figure 4) with the increase of sediment depth is observed. For the nitrogenous groups (amids bands at 1680 and 1540 cm^{-1}) the alteration process is very fast, whereas the oxygen groups (stretching of C–O at 1040 cm^{-1} and stretching of C = O at 1700 cm^{-1}) disappear more slowly[20].

Elemental analysis, FTIR and ^{13}C-NMR of the humic acids extracted from the sediments of A-zone present some differences in respect to that of B-zone, even though they maintain the peculiar characteristics of Antarctic humic acids. The extraction yields for the zone A were significantly lower. Differences are also noted for the elemental

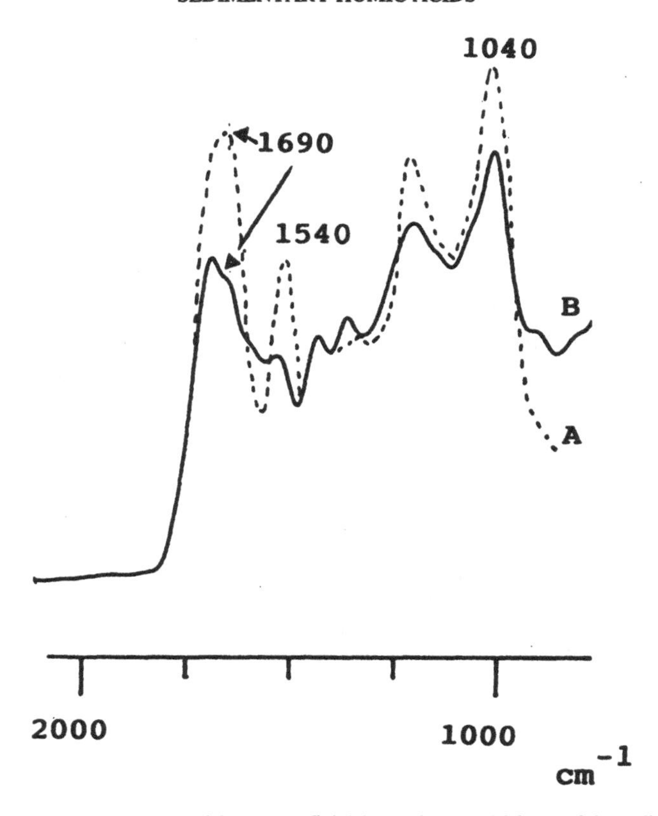

Figure 4 FT-IR spectra of HA extracted from superficial (A) and upper (B) layer of the sediment.

Table 3 Ranges of C, H, N percentages for HA (ash-free) extracted from A- and B-zone.

Zone	*Layer*	*N %*	*C %*	*H %*
A	superf.	6–7	50–60	6–7
	upper	6–7	50–60	5–6
B	superf.	7–8	70–80	7–9
	upper	6–7	50–75	7–8

composition (Table 3). Moreover HA of A-zone show, compared with those of the B-zone (Figure 5), more aromatic structures (^{13}C-NMR: zone 120–130 ppm), and major presence of carbohydrates and alcohols (FTIR: band at 1020–1159 cm^{-1}). HA of B-zone are characterized from long chains of fatty acids (peaks at 32 ppm and 2925, 2865 cm^{-1}).

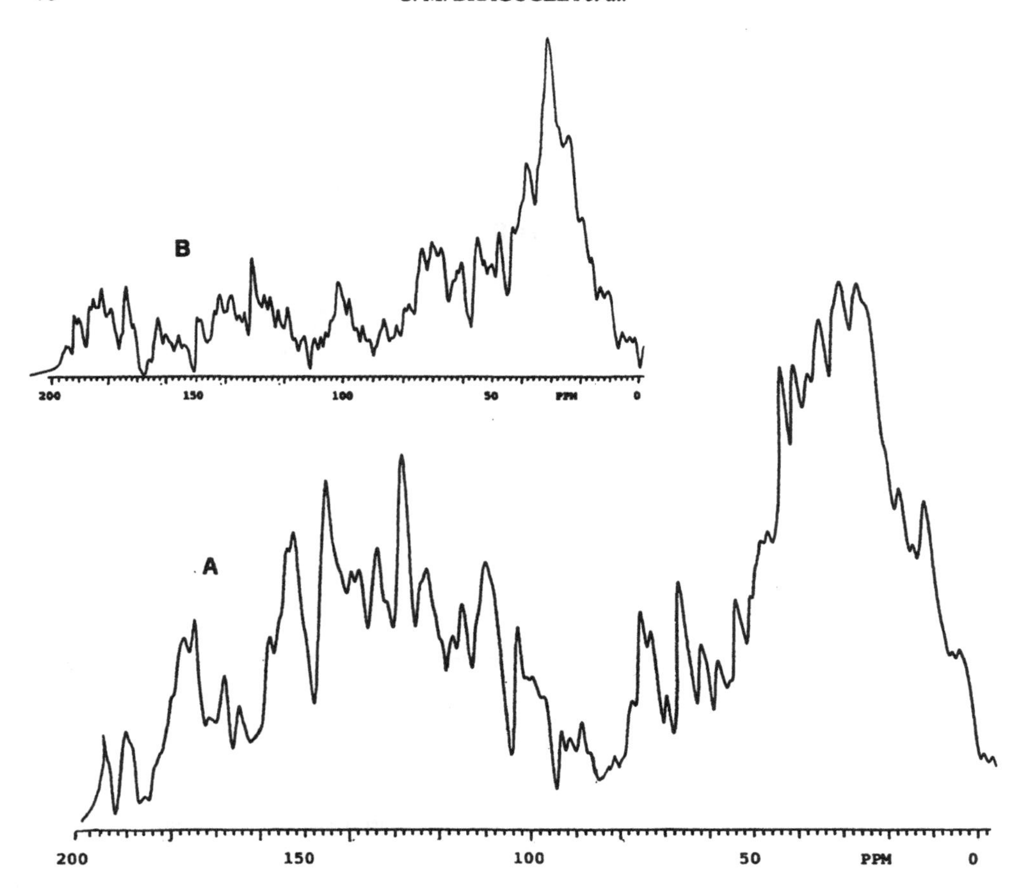

Figure 5 ^{13}C-NMR spectra of HA of the A-zone (A) and the B-zone (B).

The HA's aminoacidic content of both zones is 1–2 micromol/mg of humic acids (on ash-free basis); per cent distribution of aminoacids is reported by the hystogramms in Figure 6.

Glycine, aspartic and glutammic acid, alanine and lysine are the more abundant aminoacids (70% of the total content) in the zone A HA, whereas, in the zone B, they are glycine, aspartic and glutammic acid, alanine and serine (60% of the total content). In the zone A, lysine and glutammic acid have the same weight percentage, whereas in the zone B, the glutammic acid is present in larger amount (10–20% and 4–5%, respectively) according to the literature data[21,22]. In the zone B the basic aminoacids (hystidine, lysine and arginine) are lower than those ones of the zone A.

The differences noted for the two zones cannot derive from precursors (both autochtonous and allochtonous); different humification and sedimentary processes may justify these outcomes (an opposite temperature trend of the water column has been observed).

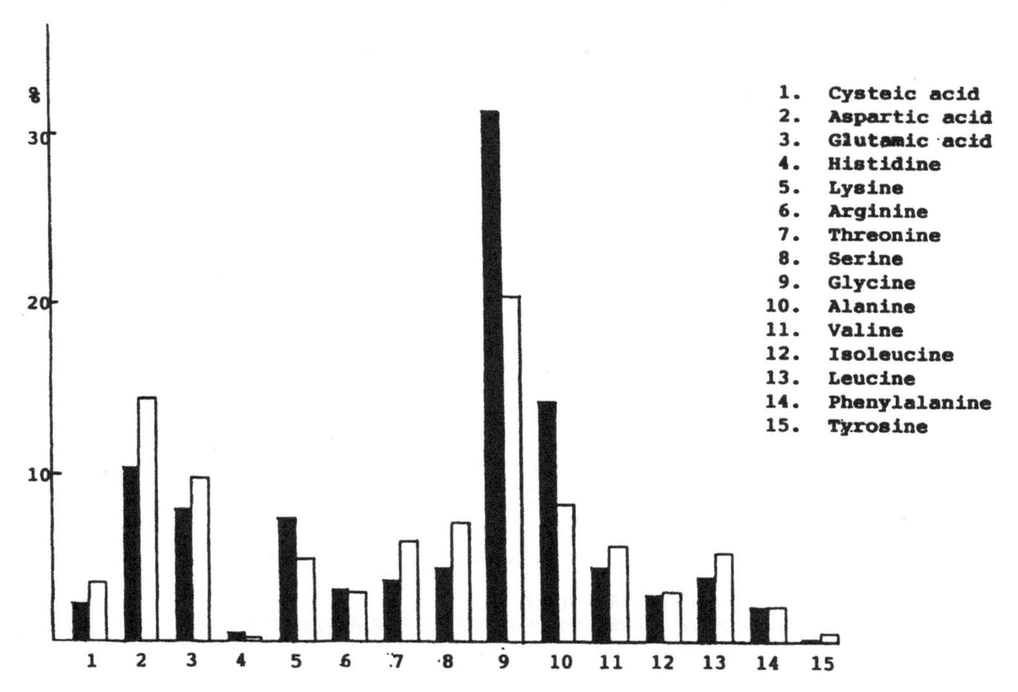

Figure 6 Aminoacid % distribution for HA of the A-zone (black coloured) and of the B-zone (white coloured).

CONCLUSIONS

The HA of marine sediments are characterized by structures that differ from the classic ones because they are affected both by the peculiar nature of precursors and the particular severe climatic conditions of Antarctica. In particular, the structures of Antarctic humic acids are generally less aromatic and their elemental composition presents a higher nitrogen and lower oxygen content.

However, some structural differences for Antarctic sedimentary HA originating from different zones have been noted, probably due to dissimilar environmental conditions (temperature, salinity, sediment depth) in which they have been formed. The observed diversities may be responsable for different complexing (and consequently different metal speciation in the sediments) and sorption capacities.

Acknowledgements

The present work was realized with the financial support of E.N.E.A. (Rome) under the Italian Antarctic Project, section 2d.3: Environmental Contamination. The authors thank Francesco Dianetti and Lucantonio Petrilli (Italian National Council of Research, Monte Libretti, Rome) for elemental analysis, Francesco Piccioni for NMR spectra and Dott. Maria Eugenia Schininà (Department of Biochemical Science "A. Rossi Fanelli", Rome) for aminoacid analysis.

References

1. M. A. Rashid and L. H. King, *Chem. Geol.*, **7**, 37–43 (1971).
2. O. K. Bordovskiy, *Mar. Geol.*, **3**, 83–114 (1965).
3. R. Ikan, T. Dorsey and I. R. Kaplan, *Anal. Chim. Acta*, **232**, 11–18 (1990).
4. G. R. Harvey and D. A. Boran, *Mar. Chem.*, **12**, 119–132 (1983).
5. G. Spezie and S. Tucci, *Annali Facolta' di Scienze Nautiche*, Volume LX (1993).
6. R. L. Wershaw and D. J. Pinkney, *Anal. Chim. Acta*, **232**, 31–42 (1990).
7. M. A. Rashid and L. H. King, *Geochim. Cosmochim. Acta*, **33**, 147–151 (1968).
8. L. Campanella, T. Ferri and B. M. Petronio, *Ann. Chim.* (Rome), **81**, 477–490 (1991).
9. L. Campanella, B. Cosma, N. Degli Innocenti, T. Ferri, B. M. Petronio and A. Pupella, *Intern. J. Environ. Anal. Chem.*, **55**, 61–75 (1994).
10. A. Nissenbaum and I. R. Kaplan, *Limnol. Ocean.*, **17**, 570–582 (1972).
11. M. A. Rashid and L. H. King, *Geochim. Cosmochim. Acta*, **34**, 193–201 (1970).
12. R. Ishiwatari, *Chem. Geol.*, **12**, 113–126 (1973).
13. B. M. Petronio, L. Campanella, N. Calace, N. Degli Innocenti and S. Tucci, *Ann. Chim.* (Rome), **84**, 81–94 (1994).
14. R. Ishiwatari, *Geochem. J.*, **1**, 61–70 (1967).
15. D. F. Cameron and M. L. Sohn, *Sci. Total Environ.*, **113**, 121–132 (1992).
16. R. Ishiwatari, Centre for Agric. Public. and Document, Wageningen, 109 (1975).
17. E. L. Poutanen and R. J. Morris, *Mar. Chem.*, **17**, 115–126 (1985).
18. S. V. Bruyevich, *Tr. Inst. Okeanol.*, Akad. Nauk S.S.S.R., 17 (1956).
19. F. S. Brown, M. J. Baedecker, A. Nissenbaum and I. R. Kaplan, *Geochim. Cosmochim. Acta*, **36**, 1185–1203 (1972).
20. A. Y. Huc and B. M. Durand, *Fuel*, **56**, 73–80 (1977).
21. G. Fengler, E. T. K. Haupt and G. Liebezeit, *Sci. Total Environ.*, **81/82**, 335–342 (1989).
22. S. Yamamoto and R. Ishiwatari, *Sci. Total Environ*, **117/118**, 279–292 (1992).

ORGANIC COMPOUNDS IN ANTARCTIC SEA-WATER AND PACK-ICE

P. G. DESIDERI, L. LEPRI, L. CHECCHINI, D. SANTIANNI, F. MASI and M. BAO

Department of Public Health, Epidemiology and Environmental Analytical Chemistry, University of Florence, Via G. Capponi 9, 50121 Florence, Italy

Pack-ice and sea-water samples collected at different depths from Terra Nova Bay and Ross Sea, during 1990/1991 Italian Antarctic Expedition, were analyzed using HRGC and GC-MS. Several classes of biogenic and anthropogenic organic compounds were identified and measured in both matrices. The results showed the changes in the organic composition at varying depths of pack-ice and sea-water and the enrichment of organic compounds in the pack.

KEY WORDS: Antarctica, pack-ice, seawater, organic pollutants, chromatographic analysis.

INTRODUCTION

The analysis of samples of pack-ice and of the immediately underlying sea-water, taken from Terra Nova Bay during the 1988/1989 Italian Antarctic Expedition, revealed the presence of numerous organic compounds in both matrices[1]. Only the sea-water surface in direct contact with the pack-ice was taken and the pack samples were cored and stored without any sectioning.

The aim of this study was to identify and measure the organic compounds present in: 1) pack-ice samples collected from different areas (Terra Nova and Wood Bay), (Figure 1-B, D and E) and in cores of 1 m, taken at different depths down to 3 m and stored separately; 2) sea-water samples taken at different depths under the pack-ice in Terra Nova Bay; 3) sea-water samples from Ross Sea collected at considerable distance from the coast (Figure 1-A) at different depths (20, 500, 1500 meters).

The results may contribute to understand the behaviour of organic compounds in Antarctic waters and during the ice formation.

EXPERIMENTAL

Sampling

The sampling stations for pack-ice and sea-water were the following (Figure 1):

Station B: Lat. 74°40' S; Long 164°07' E.
Station D: Lat. 74°36' S; Long 164°35' E.

Station E: Lat. 74°21' S; Long 165°14' E.
Station A: Lat. 70°53' S; Long 177°21' E.

The following samples were collected: pack-ice, B-1 (depth 1 m), B-2 (depth 2 m), D-1 (depth 1 m), D-2 (depth 2 m), D-3 (depth 3 m), E-1 (depth 1 m), E-2 (depth 2 m); sea-water, SWA (depth 20 m, 500 m and 1500 m), SWB (depth 0.5 m, 25 m, 250 m), SWE (depth 0.5 m).

The pack-ice samples were collected by a manually operated steel corer, after eliminating the top layer of snow. The cores (100 × 10 cm) were immediately placed in steel containers. The sea-water samples were collected in 30 liter "go-flow" bottles made of PVC with a Teflon coating. Each sample was transferred into 25 liter steel reservoirs, frozen and kept at –30°C until the time of analysis.

Reagents and materials

25 Liter stainless steel reservoirs (Inox Sabat, Bologna, Italy) were used for storage of sea-water samples. Stainless steel cylinders (1.20 × 0.15 m) were used to store the pack-ice cores. Solvents (n-hexane, methylene chloride, chloroform, acetone) were all pesticide grade purchased from Merck (GFR). Standard organic compounds are commercially available from Supelco (USA) and Alltech (USA).

Anhydrous sodium sulphate was heated for 12 hours at 450°C to remove any organic matter and then kept at 120°C until use. All apparatus was cleaned before use by repeatedly washing with chromic and concentrated sulphuric acid mixture, bidistilled water, acetone and n-hexane.

Extraction of organic compounds

The extraction of the organic compounds from the sea-water samples was performed by the replicated extractant enrichment method, which is very suitable for environmental

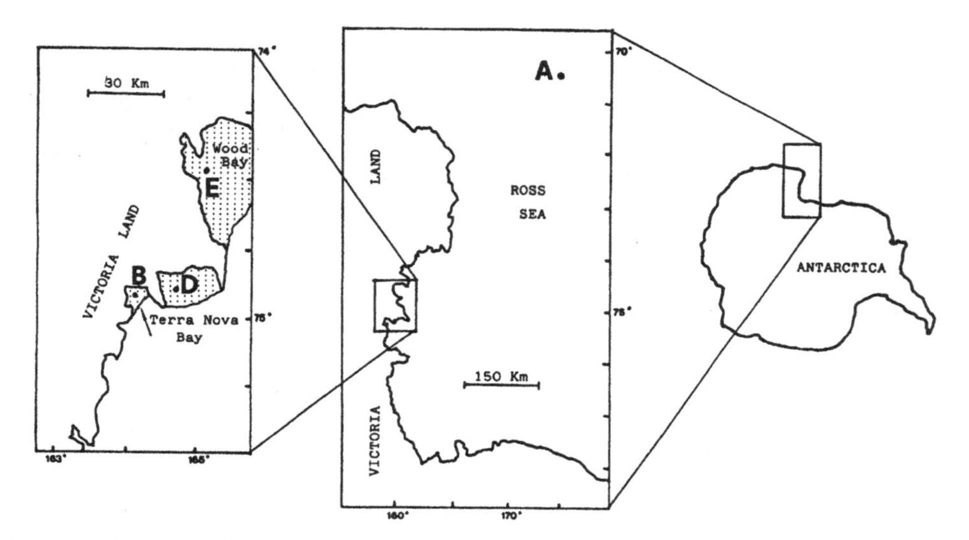

Figure 1 Sampling stations for offshore sea-water (Ross Sea, Antarctica) (A) and for pack-ice and sea-water under the pack (B, D and E).

matrices with low concentration of organic compounds, such as the Antarctic matrices. The extraction was performed by adding 3 ml of n-hexane to 3 liters of sea-water sample and then stirring the resulting mixture magnetically for 10 minutes in a special glass apparatus[2]. After leaving the mixture to stand for 5 min. the n-hexane phase was collected into a microburette (first extract). A second extraction of the same sample was performed with 1 ml of n-hexane and the resulting extract is added to the first. The enrichment of the extractant was achieved by using the n-hexane recovered from the above mentioned extraction of 3 liters of sea-water to treat equal portions of the same sea-water sample up to 9 liters[2]. This procedure was used on a total of eighteen liters of unfiltered sea-water divided in two fractions of nine liters each. Both extracts were combined, dried under sodium sulphate and cold evaporated to 100 μl under nitrogen flow in standardized conditions. The pack-ice cores, melted in a glass column at room temperature and under nitrogen flow, resulted in about 5 liters of solution. The liquid was extracted with 3 ml n-hexane following the above mentioned method and the extract was treated as above.

Fractionaction of the organic extracts

The fractionaction of the organic extracts of both matrices was performed on a silicagel 60 HR for TLC (Merck) column (0.6 × 10 cm) previously activated to 120°C for 12 hours. The n-hexane extract (100 μl) was deposited at the top of the column and the organic compounds were fractionated using the following eluents:

1) 5 ml of n-hexane: n-alkanes
2) 5 ml of n-hexane/toluene 9.4/5.6 v/v: polycyclic aromatic hydrocarbons
3) 5 ml of n-hexane/ethyl acetate 9/1 v/v: aldehydes, ketones, phthalates, fatty acid esters
4) 5 ml of n-hexane/ethyl acetate 6/4 v/v: alcohols, phthalates

The volume of each fraction was cold evaporated to 100 μl under nitrogen flow.

The entire analytical procedure was repeated five times for every sample.

The recoveries obtained for the different classes of compounds, calculated by using a standard mixture containing n-alkanes, polycyclic aromatic hydrocarbons, aliphatic aldehydes, ketones, alcohols, fatty acid esters and phthalates, at concentration levels of 10 ng/l, are indicated in Table 1.

Derivatization of aldehydes, ketones and alcohols

The fraction containing aldehydes and ketones was derivatized by using O-(2,3,4,5,6-pentafluorobenzyl)-hydroxylamine hydrochloride (PFBHA) to obtain the corresponding

Table 1 Recoveries of organic compounds from 9 liters of an aqueous standard solution.

Organic compounds	*Recovery (%)*	*St. dev.*
n-Alkanes	85	5
PAHs	80	7
Aldehydes, ketones, alcohols	75	5
Phthalates	70	10

oximes (PFBO)[3]. The alcohols were transformed into their trimethylsilyl derivatives by treatment with N,O-bis(trimethylsilyl)trifluoroacetamide (BSTFA)[4].

HRGC and GC-MS analysis

For the determination of organic compounds, a HRGC-5160 Mega Series (Carlo Erba, Italy) gas chromatograph equipped with a FID detector was used. The injection was made by using a Cold-SSL injector (Carlo Erba) according to the following temperature program: injection at 40°C, then a rapid increase in temperature to 300°C and splitting after 30 sec.. Column temperature program: 40°C for 1 min., then linear increase to 300°C at 4°C/min., and finally isotherm at 300°C for 15 min.. Capillary columns Supelco PTE-5: (30 m, 0.25 mm i.d., 0.25 µm thickness). Carrier gas:helium. The chromatographic peaks were analyzed with a Mega-2 computer system (Carlo Erba) with Spectra Physics software. Confirmatory GC-MS analyses of aldehydes, ketones, alcohols and esters were performed on a Varian 3400 gas chromatograph coupled with a Finnigan ITD mass detector. Carrier gas:helium. An injector SPI (Varian) was used according to the following temperature program: injection at 40°C, then a rapid increase to 300°C. The column temperature program was the same as described above.

Identification of organic compounds

The identification and quantitative determination of organic compounds was realized using gas-chromatographic retention indices with 8 n-alkanes (C-8, C-12, C-16, C-20, C-24, C-28, C-32 and C-34) as standards and a reference calibration table and/or comparing their mass spectra with those reported in the N.B.S. library and a second library made in our laboratory on ITD.

The exact name is reported in the Tables only for those compounds positively identified with the above methods, while only the belonging class is given for the others.

RESULTS AND DISCUSSION

Pack-ice

Tables 2 and 3 show the organic compounds identified in the three pack-ice samples (B, D, E) with their concentrations. From samples B and E only two cores corresponding to depths of one (B-1, E-1) and two meters (B-2, E-2) were obtained while three cores of one meter each were taken in the same place for sample D. The results reveal that this Antarctic matrix is quite rich in both biogenic and anthropogenic organic compounds as was found in pack-ice samples taken from Terra Nova Bay during the Italian Antarctic Expediction 1988/89[1].

The distribution of biogenic compounds, such as aldehydes, ketones, alcohols and squalene is rather homogeneous in the three samples and indicates the validity of the obtained results. The alcohols exhibited the higher concentrations among the different classes of organic compounds. In a number of pack-ice cores squalene dominated the aliphatic hydrocarbon fraction, i.e. 62%–78% of the aliphatic fraction in D-2 and E-2 cores (Table 2).

Table 2 Aliphatic hydrocarbons in Antarctic pack-ice (ng/l); medium values of five determinations with standard deviation.

Compounds	*B-1* *1 m*	*B-2* *2 m*	*D-1* *1 m*	*D-2* *2 m*	*D-3* *3 m*	*E-1* *1 m*	*E-2* *2 m*
n-ALKANES							
n-C 14	bdl	2 *	6 ± 2	2 *	4 ± 1	2 *	2 *
n-C 15	bdl	bdl	5 ± 1	bdl	8 ± 2	3 ± 1	7 ± 2
n-C 16	2 *	8 ± 2	6 ± 2	9 ± 2	15 ± 4	26 ± 7	6 ± 2
n-C 17	2 *	3 ± 1	3 ± 1	8 ± 2	16 ± 4	5 ± 2	6 ± 2
n-C 18	2 *	2 *	3 ± 1	bdl	15 ± 4	6 ± 2	8 ± 2
n-C 19	2 *	bdl	2 *	bdl	12 ± 3	2 *	10 ± 3
n-C 20	bdl	5 ± 2	2 *	2 *	16 ± 4	5 ± 1	9 ± 2
n-C 21	5 ± 1	4 ± 1	7 ± 2	2 *	26 ± 6	10 ± 3	11 ± 2
n-C 22	3 ± 1	4 ± 1	4 ± 1	2 *	32 ± 6	11 ± 3	10 ± 2
n-C 23	4 ± 1	6 ± 2	5 ± 2	2 *	29 ± 5	7 ± 2	6 ± 1
n-C 24	4 ± 1	4 ± 1	21 ± 4	4 ± 1	42 ± 6	11 ± 2	10 ± 2
n-C 25	7 ± 2	10 ± 3	7 ± 2	2 *	30 ± 4	12 ± 3	6 ± 1
n-C 26	8 ± 2	3 ± 1	9 ± 3	2 *	27 ± 4	13 ± 3	6 ± 1
n-C 27	8 ± 2	2 *	7 ± 2	2 *	22 ± 4	10 ± 4	6 ± 1
n-C 28	10 ± 3	3 ± 1	8 ± 3	2 *	27 ± 5	9 ± 3	8 ± 2
n-C 29	9 ± 3	bdl	6 ± 2	2 *	33 ± 7	8 ± 3	5 ± 1
n-C 30	5 ± 2	bdl	4 ± 1	bdl	13 ± 3	5 ± 2	2 *
n-C 31	3 ± 1	bdl	2 *	bdl	18 ± 4	5 ± 1	4 ± 1
n-C 32	bdl	bdl	2 *	bdl	11 ± 3	2 *	2 *
Total n-alkanes	74	56	109	41	396	152	124
Odd/even predominance	1.17	0.86	0.75	0.86	0.98	0.70	1.00
Squalane	bdl	bdl	bdl	bdl	8 ± 1	bdl	bdl
Squalene	61 ± 13	41 ± 11	68 ± 18	67 ± 17	540 ± 135	39 ± 9	431 ± 86

* = detection limit; bdl = below detection limit.

The n-alkanes were quantified over the range C-14 to C-32 in accordance with Cripps[5] (C-15 to C-30) and Green *et al.*[6] (C-15 to C-33); the concentration range was between 41–396 ng/l (Table 2) and agrees with the data reported in previous studies[1,7]. Similar levels of n-alkanes (70–170 ng/l) were found by Green *et al.*[6] in sea-water samples taken at the Davis Station (Eastern Antarctica) but lower (4.1–10.4 ng/l) and higher (1.1–21.8 μg/l) levels were reported by Sanchez-Pardo *et al.*[8] and Cripps[9] respectively for the same area of Antarctica (Bransfield Strait). This variation in the concentration is probably related to complex combination of biological properties of the Southern Ocean.

The odd to even carbon number ratio for n-alkanes gave doubtful informations on their origin[10]. This ratio, evaluated in the range nC-15 to nC-32, was greater than unity for core B-1, near to unity for cores D-3 and E-2 and less than one for the others (B-2, E-1, D-1, D-2) (Table 2). In addition, the values for the cores B-2 and D-2 were not very significant since the concentrations of most alkanes are equal to or less than the detection limit.

Odd to even ratio with values in the range 0.7–1.4 were found on a number of pelagic species from the Bransfield Strait[10] but several authors[11,12] reported that even numbered n-alkanes predominated in Antarctic marine organism from Ross Sea and, therefore the n-alkanes in cores B-2, E-1, D-1 and D-2 are probably biogenic.

Table 3 Organic etherocompounds in Antarctic pack-ice (ng/l); medium values of five determinations with standard deviation.

Compounds	*B-1 1 m*	*B-2 2 m*	*D-1 1 m*	*D-2 2 m*	*D-3 3 m*	*E-1 1 m*	*E-2 2 m*
ALDEHYDES							
Nonanal	36 ± 9	69 ± 17	72 ± 22	49 ± 12	47 ± 15	27 ± 8	46 ± 14
Aliphatic aldehyde	3 ± 1	10 ± 3	30 ± 8	bdl	17 ± 5	23 ± 6	18 ± 5
Decanal	14 ± 4	23 ± 7	17 ± 4	19 ± 4	18 ± 4	12 ± 3	20 ± 5
Decenal	12 ± 3	12 ± 3	49 ± 10	22 ± 7	20 ± 5	11 ± 2	11 ± 3
Undecanal	7 ± 2	10 ± 2	8 ± 2	5 ± 1	11 ± 2	bdl	11 ± 2
Dodecanal	bdl	6 ± 1	3 ± 1	3 ± 1	3 ± 1	5 ± 2	6 ± 2
Tridecanal	7 ± 2	4 ± 1	2 *	3 ± 1	bdl	bdl	bdl
Tetradecanal	bdl	bdl	bdl	bdl	3 ± 1	2 *	3 ± 1
Total aliphatic aldehydes	79	134	181	101	119	80	115
KETONES							
2-Nonanone	2 *	9 ± 3	3 ± 1	4 ± 1	bdl	2 *	4 ± 1
Aliphatic ketone	2 *	5 ± 2	5 ± 1	38 ± 11	2 *	2 *	5 ± 2
Aliphatic ketone	bdl	5 ± 2	7 ± 2	3 ± 1	3 ± 1	bdl	4 ± 1
2-Decanone	6 ± 2	11 ± 3	15 ± 4	6 ± 2	bdl	3 ± 1	bdl
Aliphatic ketone	8 ± 2	42 ± 10	20 ± 5	59 ± 18	35 ± 10	22 ± 6	4 ± 1
2-Undecanone	5 ± 2	10 ± 2	4 ± 1	5 ± 2	3 ± 1	2 *	2 *
Aliphatic ketone	9 ± 3	5 ± 1	6 ± 2	3 ± 1	3 ± 1	10 ± 3	6 ± 1
2-Dodecanone	12 ± 3	21 ± 6	14 ± 4	7 ± 2	3 ± 1	13 ± 3	12 ± 3
2-Tridecanone	6 ± 2	4 ± 1	3 ± 1	3 ± 1	bdl	bdl	bdl
2-Tetradecanone	6 ± 2	2 *	12 ± 4	4 ± 1	2 *	11 ± 3	3 ± 1
Aliphatic ketone	3 ± 1	3 ± 1	bdl	8 ± 3	bdl	bdl	bdl
Total aliphatic ketones	56	114	89	132	51	65	40
ALCOHOLS							
1-Docosanol	49 ± 12	89 + 20	61 ± 14	92 ± 19	132 ± 37	66 ± 14	113 ± 23
1-Tetradecanol	29 ± 6	44 ± 9	25 ± 6	17 ± 4	15 ± 5	45 ± 10	16 ± 4
1-Hexadecanol	27 ± 7	32 ± 9	14 ± 4	42 ± 11	16 ± 5	25 ± 6	30 ± 8
1-Octadecanol	59 ± 18	24 ± 7	27 ± 8	57 ± 17	90 ± 24	41 ± 11	89 ± 22
1-Eicosanol	22 ± 7	13 ± 4	20 ± 6	12 ± 3	61 ± 17	51 ± 12	34 ± 11
Total aliphatic alcohols	186	202	147	220	314	228	282
PHTHALATES							
Di-iso-butylphthalate	248 ± 32	299 ± 36	495 ± 64	199 ± 22	361 ± 54	815 ± 98	283 ± 45
Di-n-butylphthalate	311 ± 56	123 ± 21	268 ± 46	142 ± 21	572 ± 63	351 ± 32	174 ± 21
Benzylbutylphthalate	1056 ± 84	3134 ± 345	236 ± 17	113 ± 11	152 ± 21	421 ± 21	107 ± 10
Bis(2-ethylhexyl)phthalate	54 ± 9	33 ± 3	196 ± 29	67 ± 6	231 ± 30	25 ± 3	60 ± 7
Total phthalates	1669	3589	1195	521	1316	1612	624

* = detection limit; bdl = below detection limit.

Further informations can be deduced from the n-alkane profiles for the more representative ice-cores. The hydrocarbon profile for core B-1 was dominated by nC23–nC30 alkanes and suggested an input from terrestrial vegetation in which odd carbon n-alkanes in the range nC25–nC33, derived from cuticular waxes of continental plants, are dominant[13]. Consequently, the continental contribution defines the n-alkane fingerprinting and the odd/even ratio in the core B-1. The n-alkane profiles of the other cores, on the contrary, do not show any predominance of nC23–nC30 compounds. There is, however, evidence of specific species influencing the hydrocarbon composition (17% nC16 and 19% nC24 in cores E-1 and D-1 respectively). The dominance of one alkane in one seawater sample (14% nC17) and in some sediments (44% nC18) was already observed by Green *et al.*[6] and it is probably due to a particular algal or bacterial species[14] and, therefore, to biogenic local sources.

The trend of the concentration of biogenic compounds for the three samples as the depth changes is shown in Figure 2.

The alcohols have a uniform behaviour, since increase in all samples with depth. It is not possible to determine any specific trend for the two other classes of organic compounds. It is important to note that the content of the various classes generally corresponds to the one reported for the same compounds in the Antarctic environmental matrices. In fact the concentration range is between 78–181 ng/l for aldehydes, 40–132 ng/l for ketones, 147–314 ng/l for alcohols and agrees with the data reported in previous studies[1,7].

The range of phthalate concentrations was between 521–1669 ng/l with the exception of 3589 ng/l in core B-2 (Table 3). These values depend on the high levels of benzylbutylphthalate in all cores and, particularly, in core B-2 which is the nearest to the Italian Base. Consequently, the presence of benzylbutylphthalate, which was not found in previous pack-ice samples[1], is probably characteristic of a local anthropogenic origin. The other three phthalates (di-n-butyl-, di-iso-butyl-, di-2-ethylhexyl-), on the contrary, are ubiquitous environmental contaminants and are present in all Antarctic matrices examined up to now (seawater, pack-ice, snow, sediments). The concentration range is 408–1191 ng/l (Table 3), similar levels of these compounds were found in pack-ice (525–745 ng/l) and seawater (139–556 ng/l) samples taken during the Expedition 1988/89[1]. The range of the phthalate concentrations in the North Sea (Liverpool Bay) is 467–5840 ng/l near the Tees and Mersey estuaries[15]. These data demonstrate the pollution level of Terra Nova Bay. A source for phthalates is the long range atmospheric transport as shown by the high concentrations of these compounds in the snow[7] and in the particulate organic matter sampled at Terra Nova Bay Station[16]. This fact suggestes that areosol transport is the dominant mechanism for input of phthalates to Antarctic precipitation. Local human activity in Ross Sea and on land may even result in phthalates contamination of the marine environment.

Sea-water under the pack and offshore

Tables 4 and 5 show the concentrations of organic compounds found in four sea-water samples collected under the pack in two different areas, B and E. The latter area is the furthest from the Italian Base. The same Tables also report the organic substances identified in the offshore sea-water samples (area A) taken at depths of 20 m (SWA-20), 500 m (SWA-500) and 1500 m (SWA-1500).

In all the samples the following compounds were identified: n-alkanes, aldehydes, ketones and alcohols. High concentrations of phthalates were also found but such data have not been reported because the samples were contaminated by the "go-flow" bottles.

The range of the n-alkane concentrations is 36–260 ng/l (Table 4), and the hydrocarbon level is generally lower than the one found for the pack-ice cores. Almost all n-alkanes were quantified since their detection limit is lower in water (1 ng/l) than in pack-ice (2 ng/1) owing to the greater volumes of seawater samples.

The odd/even predominance for the n-alkanes is much lower than 1 in all samples indicating that their origin is biogenic. It should be noted that Green *et al.*[6] found

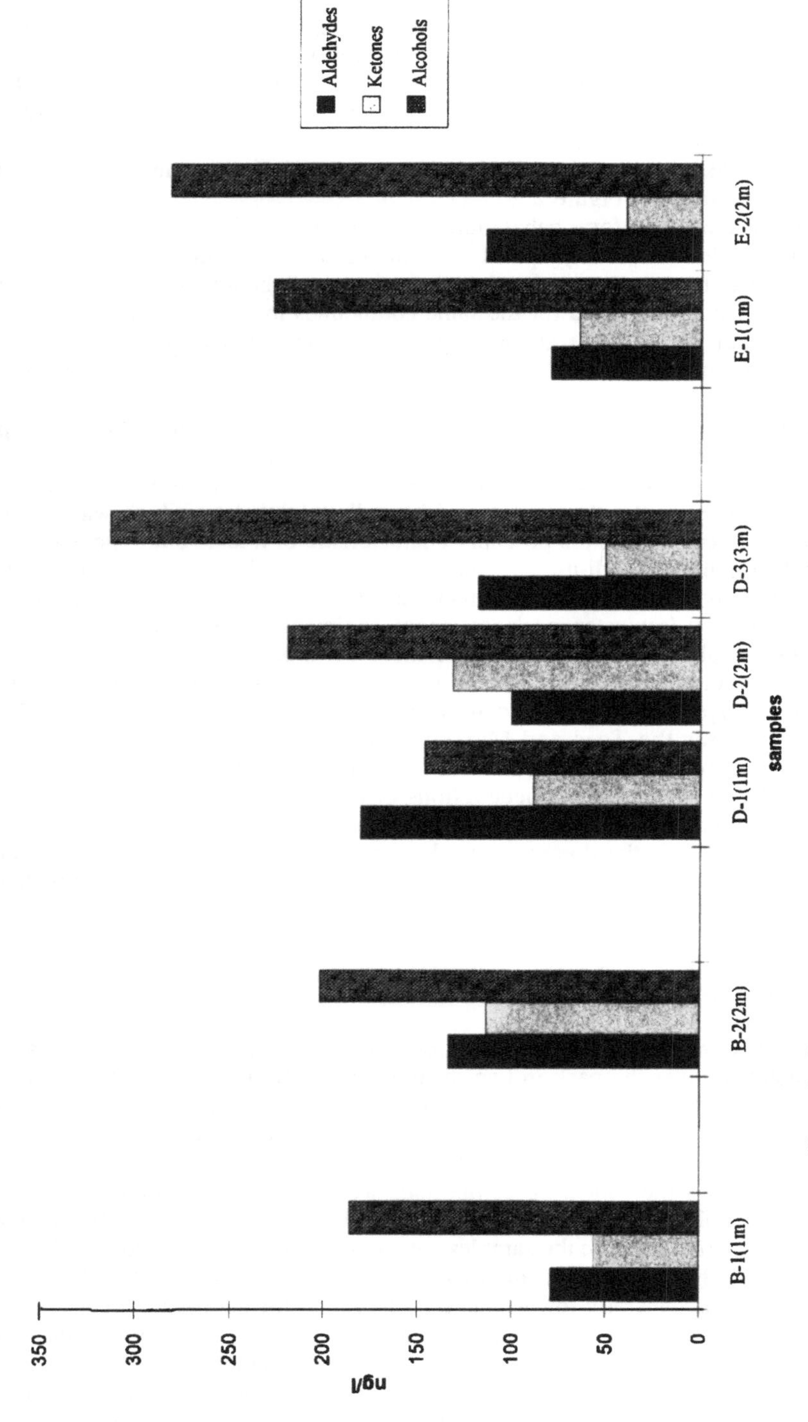

Figure 2 Concentrations (ng/l) of classes of biogenic organic compounds in pack-ice samples collected from different areas.

Table 4 Aliphatic hydrocarbons in offshore sea-water samples and in sea-water samples collected under pack-ice (ng/l); medium values of five determinations with standard deviation.

	SW samples collected under pack-ice				*Offshore SW samples*		
Compounds	*SWB 0.5 m*	*SWB 25 m*	*SWB 250 m*	*SWE 0.5 m*	*SWA 20 m*	*SWA 500 m*	*SWA 1500 m*
n-ALKANES							
n-C 14	1 *	1 *	1 *	1 *	5 ± 1.6	1 *	1 *
n-C 15	1 *	bdl	1 *	1 *	7 ± 1.9	2 ± 0.5	1 *
n-C 16	8 ± 1.8	6 ± 1.3	8 ± 1.9	7 ± 1.8	24 ± 7.0	14 ± 3.4	14 ± 3.6
n-C 17	3 ± 0.6	2 ± 0.4	2 ± 0.4	2 ± 0.5	12 ± 3.1	3 ± 0.7	bdl
n-C 18	2 ± 0.4	1 *	1 *	2 ± 0.5	10 ± 2.3	2 ± 0.4	1 *
n-C 19	2 ± 0.4	bdl	1 *	2 ± 0.4	10 ± 2.2	1 *	2 ± 0.4
n-C 20	1 *	1 *	1 *	1 *	13 ± 2.6	2 ± 0.4	2 ± 0.4
n-C 21	2 ± 0.3	1 *	2 ± 0.3	2 ± 0.4	13 ± 2.2	2 ± 0.4	2 ± 0.3
n-C 22	2 ± 0.2	1 *	1 *	2 ± 0.3	8 ± 1.4	1 *	2 ± 0.3
n-C 23	3 ± 0.2	1 *	1 *	3 ± 0.4	8 ± 1.2	1 *	3 ± 0.4
n-C 24	11 ± 1.3	8 ± 1.0	9 ± 1.3	7 ± 1.1	34 ± 4.4	10 ± 1.1	17 ± 2.2
n-C 25	4 ± 0.4	1 *	2 ± 0.3	4 ± 0.6	11 ± 1.2	2 ± 0.2	5 ± 0.5
n-C 26	4 ± 0.4	1 *	4 ± 0.4	3 ± 0.4	16 ± 2.4	2 ± 0.3	6 ± 0.8
n-C 27	3 ± 0.3	1 *	2 ± 0.2	3 ± 0.5	10 ± 1.4	2 ± 0.3	4 ± 0.6
n-C 28	8 ± 0.9	7 ± 1.0	8 ± 1.2	5 ± 0.8	22 ± 3.7	7 ± 1.1	13 ± 2.3
n-C 29	3 ± 0.4	1 *	2 ± 0.3	3 ± 0.6	10 ± 1.6	2 ± 0.4	3 ± 0.7
n-C 30	2 ± 0.4	bdl	1 *	1 *	13 ± 2.7	1 *	2 ± 0.5
n-C 31	2 ± 0.4	1 *	bdl	1 *	7 ± 1.7	1 *	1 *
n-C 32	7 ± 1.7	2 ± 0.4	5 ± 1.1	3 ± 0.8	27 ± 7.3	6 ± 1.5	7 ± 1.8
Total n-alkanes	69	36	52	53	260	62	86
Odd/even predominance	0.51	0.30	0.34	0.68	0.53	0.36	0.33

* = detection limit; bdl = below detection limit.

odd/even ratios between 0.4–0.8 for in Antarctic seawater particulate. The hydrocarbon profiles were similar and the alkanes nC16, nC24 and nC28 were dominant in all seawater samples. These data suggest a biogenic origin too.

Aliphatic aldehydes and ketones have a number of carbon atoms less than sixteen while the alcohols contain up to twenty (Table 5). The results indicate that there are no appreciable differences in the composition of organic compounds found in sea-water samples collected under the pack or offshore (Ross Sea). Figure 3 shows the trend of the concentration for the biogenic organic compounds at varying depths. The concentrations of the substances in the samples collected under the pack in two different areas (SWB and SWE) and at the same depth (0.5 m) are similar with the exception of the alcohols. The samples taken from area B at 0.5 m, 25 m and 250 m, contain the same quantities of biogenic compounds but different amounts of ketones.

Comparison of pack-ice and seawater

To obtain information about the possible enrichment of organic compounds in the pack, we compared the concentration range of each class of substances in the pack, in the seawater under the pack and in offshore seawater (see Figure 4). The results show that

Table 5 Organic etherocompounds in offshore sea-water samples and in sea-water samples collected under pack-ice (ng/l); medium values of five determinations with standard deviation.

	SW samples collected under pack-ice				*Offshore SW samples*		
Compounds	*SWB 0.5 m*	*SWB 25 m*	*SWB 250 m*	*SWE 0.5 m*	*SWA 20 m*	*SWA 500 m*	*SWA 1500 m*
ALDEHYDES							
Nonanal	2 ± 0.6	2 ± 0.7	4 ± 1.4	bdl	1 *	bdl	1 *
Aliphatic Aldehyde	bdl	bdl	2 ± 0.6	bdl	5 ± 1.7	2 ± 0.6	1 *
Decanal	5 ± 1.2	3 ± 0.8	2 ± 0.4	4 ± 1.0	9 ± 2.4	4 ± 1.2	5 ± 1.3
Decenal	2 ± 0.5	1 *	1 *	1 *	7 ± 2.1	1 *	1 ± 0.3
Undecanal	1 *	bdl	1 *	1 *	5 ± 1.3	2 ± 0.6	3 ± 1.0
Dodecanal	1 *	bdl	bdl	bdl	4 ± 0.9	1 *	1 *
Tridecanal	bdl	bdl	bdl	bdl	2 ± 0.4	2 ± 0.6	1 *
Tetradecanal	bdl	bdl	1 *	bdl	2 ± 0.5	1 *	1 *
Total aliphatic aldehydes	11	6	11	6	35	13	14
KETONES							
2-Nonanone	1 *	1 *	2 ± 0.6	1 *	14 ± 5.0	7 ± 2.4	4 ± 1.2
Aliphatic ketone	bdl	1 *	1 *	bdl	7 ± 2.0	2 ± 0.6	3 ± 1.0
Aliphatic ketone	2 ± 0.7	1 *	6 ± 1.7	1 *	17 ± 5.8	6 ± 2.2	5 ± 1.8
2-Decanone	1 *	2 ± 0.7	7 ± 2.0	bdl	24 ± 7.4	7 ± 2.0	12 ± 3.6
Aliphatic ketone	bdl	2 ± 0.7	3 ± 0.8	bdl	24 ± 9.1	12 ± 4.3	4 ± 1.4
2-Undecanone	8 ± 1.9	2 ± 0.6	3 ± 0.7	1 *	14 ± 3.8	3 ± 0.9	6 ± 1.6
Aliphatic ketone	bdl	2 ± 0.6	bdl	1 *	8 ± 2.5	2 ± 0.7	5 ± 1.7
2-Dodecanone	1 *	bdl	8 ± 1.6	bdl	6 ± 1.5	2 ± 0.6	1 *
2-Tridecanone	bdl	bdl	1 *	bdl	3 ± 0.8	2 ± 0.6	bdl
2-Tetradecanone	bdl	1 *	2 ± 0.4	bdl	3 ± 0.9	2 ± 0.7	bdl
Aliphatic ketone	2 ± 0.6	3 ± 0.8	2 ± 0.6	1 *	3 ± 1.0	4 ± 1.4	5 ± 1.7
Total aliphatic ketones	15	15	35	5	123	49	45
ALCOHOLS							
1-Docosanol	7 ± 2.0	4 ± 1.0	6 ± 1.6	41 ± 9.8	17 ± 4.9	36 ± 11.2	50 ± 15.5
1-Tetradecanol	11 ± 2.9	11 ± 2.9	10 ± 2.8	45 ± 9.9	35 ± 10.9	2 ± 0.6	6 ± 1.7
1-Hexadecanol	8 ± 2.2	9 ± 2.3	14 ± 4.1	10 ± 2.7	34 ± 9.2	5 ± 1.4	11 ± 3.1
1-Ottadecanol	19 ± 5.5	2 ± 0.5	18 ± 4.0	16 ± 4.5	43 ± 10.8	15 ± 3.9	10 ± 2.6
1-Eicosanol	14 ± 4.3	11 ± 2.9	7 ± 1.7	4 ± 1.2	39 ± 10.9	6 ± 1.9	3 ± 0.9
Total aliphatic alcohols	59	37	55	116	168	64	80

* = detection limit; bdl = below detection limit.

the concentration levels of each class of organic substances in the pack are consistently higher than those in seawater taken under the pack and in the offshore seawater. These data confirm previous results found for the same matrices[1] and the conclusion can be made that there is a real enrichment of organic compounds from seawater during the pack-ice formation. This agrees with the hypothesis that the pack forms from the surface down and incorporates all the organic compounds present at the marine surface microlayer which is, as known, more enriched in organic compounds than the underlying waters[17].

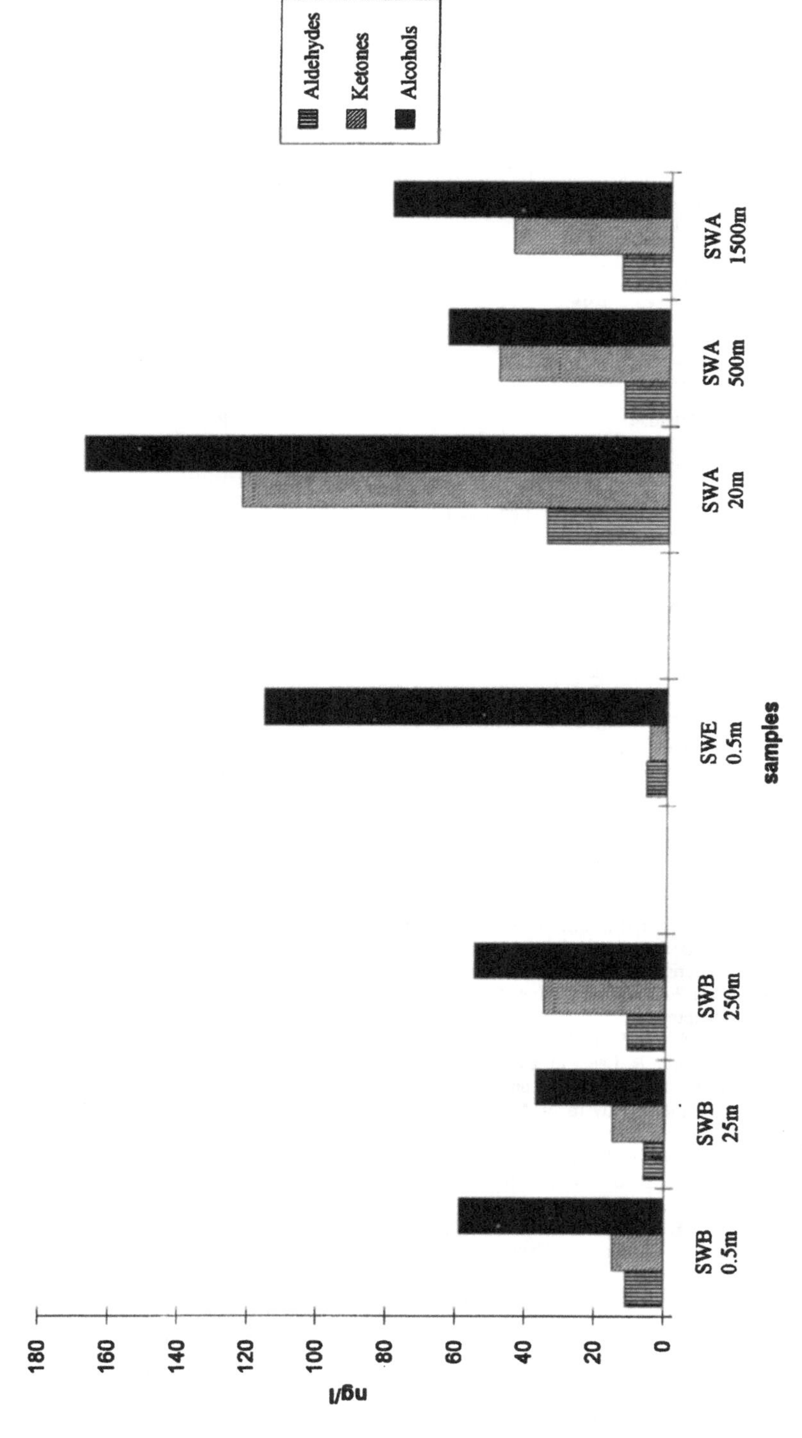

Figure 3 Concentrations (ng/l) of biogenic organic compounds in sea-water samples at different depths.

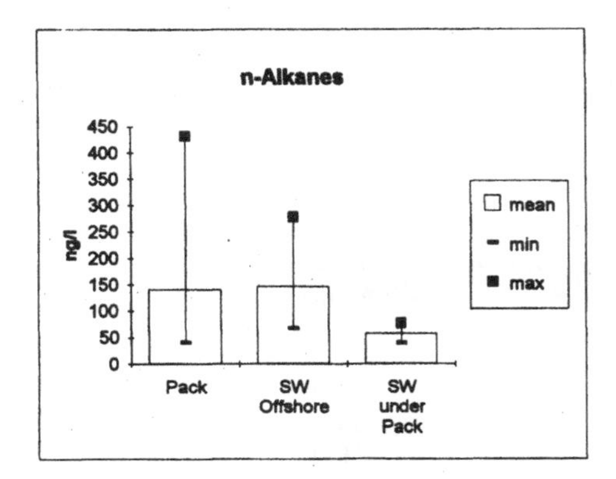

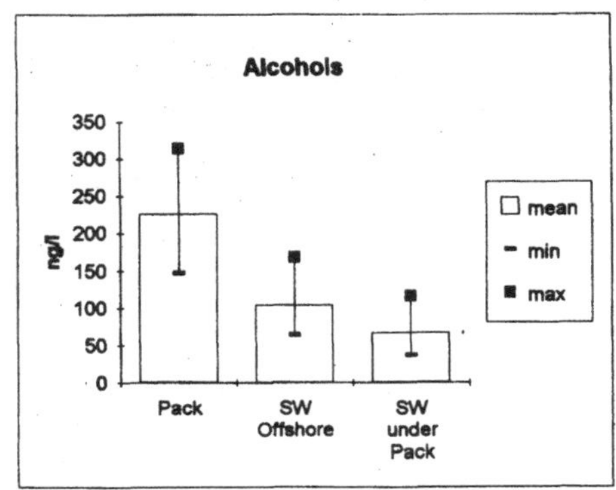

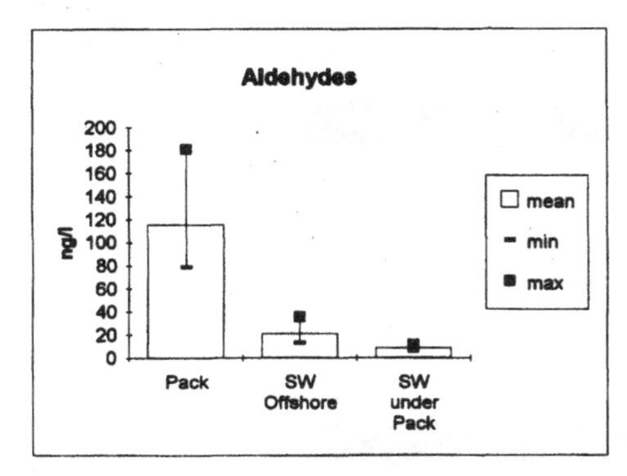

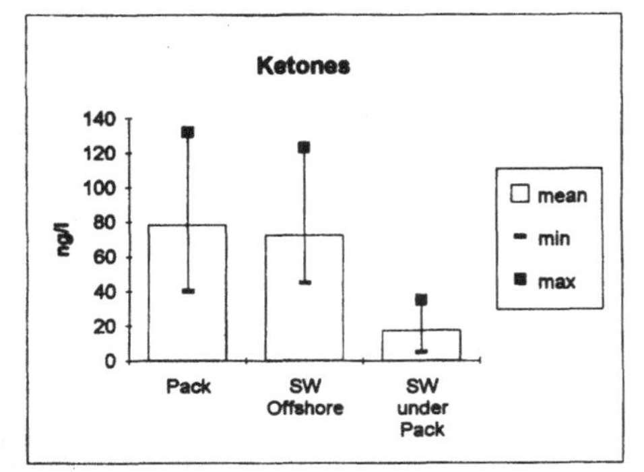

Figure 4 Comparison of concentration ranges of classes of organic compounds in pack-ice, offshore sea-water and sea-water collected under pack-ice.

References

1. P. G. Desideri, L. Lepri and L. Checchini, *Ann. Chim. (Rome)*, **81,** 395–416 (1991).
2. P. G. Desideri, L. Lepri and L. Checchini, *Mikrochim. Acta*, **107,** 55–63 (1992).
3. A. Darbre in *Handbook of Derivatives for Chromatography*, (K. Blau and G. S. King eds., Heidens and Son Ltd. 1978) chapter 2.
4. C. F. Poole in *Handbook of Derivatives for Chromatography*, (K. Blau and G. S. King eds., Heidens and Son Ltd. 1978) chapter 4.
5. G. C. Cripps, *Mar. Pollut. Bull.*, **24,** 109–114 (1992).
6. G. Green, J. H. Skerratt, R. Leeming and P. D. Nichols, *Mar. Pollut. Bull.*, **25,** 293–302 (1992).
7. P. G. Desideri, L. Lepri, L. Checchini and D. Santianni, *Intern. J. Environ. Anal. Chem.*, **55,** 33–46 (1994).
8. J. Sanchez-Pardo and J. Rovira in *Actas del Segundo Symposium Espanol de Estudios Amtarticos*, (J. Castellvi ed., Madrid, Consejo Superior de Investigaciones Cientificas 1987) pp. 117–124.
9. G. C. Cripps and J. Priddle, *Ant. Sci.*, **3,** 233–250 (1991).
10. G. C. Cripps, *Ant. Sci.*, **1,** 307–312 (1989).
11. R. J. Nachman, *Lipids*, **20,** 629–633 (1985).
12. S. B. Reinhardt and E. S. Van Vleet, *Mar. Biol.*, **91,** 149–159 (1986).
13. G. Eglinton and R. J. Hamilton in *Chemical Plant Taxonomy*, (T. Swain ed., Academic Press, 1963), pp. 187–207.
14. A. Saliot in *Marine Organic Chemistry*, (E. K. Duursma and P. Dawson eds., Elsevier, Amsterdam, 1981), pp. 327–374.
15. R. J. Lowe, T. W. Fileman and P. Matthiessen, *Water Sci. Tech.*, **24,** 127–134 (1991).
16. P. Ciccioli, A. Cecinato, E. Brancaleoni, M. Montagnoli and L. Allegrini, *Intern. J. Environ. Anal. Chem.*, **55,** 47–59 (1994).
17. J. C. Marty, V. Zutic, R. Precali, A. Saliot, B. Cosovic, N. Smodlaka and G. Cauwet, *Mar. Chem.*, **25,** 243–263 (1988).

POLYCHLOROBIPHENYLS IN SEDIMENT, SOIL AND SEA WATER SAMPLES FROM ANTARCTICA

R. FUOCO[a], M. P. COLOMBINI[a], C. ABETE[b] and S. CARIGNANI[a]

[a]Dipartimento di Chimica e Chimica Industriale dell'Universita' di Pisa Via Risorgimento 35, 56126 Pisa, Italy, [b]Istituto di Chimica Analitica Strumentale del C.N.R. via Risorgimentoo 35, 56124 Pisa, Italy

The presence of PCBs was evaluated in environmental samples collected in Antarctica during the 1990–91 and 1991–92 Italian Expeditions by using an optimized procedure for GC-ECD peak assignment. In particular, marine sediment samples from Terra Nova Bay and Ross Sea, and lake sediment and soil samples from Victoria Land, near to the Italian Base (BTN) (1990–1991 expedition) were collected and analyzed. The relevant PCB concentrations ranged between 30–160, 60–120 and 40–70 pg/g respectively, and were strongly dependent on the particle size distribution of each sample as found in previous expeditions. The depth profiles of PCB content in marine sediment samples collected in a few stations clearly show that PCBs are confined in a surface layer of about 10 cm. A coastal sea water depth profile of PCBs before and after pack ice melting was also obtained by collecting samples in Terra Nova Bay—Gerlache Inlet (1990–91 expedition). The total PCB concentration was about 140 pg/l and was practically constant up to 25 m deep. At 250 m which is near the sea bed, an increase of PCB content up to about 200 pg/l was observed. Finally, PCBs were measured in sea water samples collected in the same area (1991–92 expeditions), showing an increase of about 70% in the surface water layer after pack ice melting.

KEY WORDS: Antarctica, PCBs, sea water, marine and lake sediments, soils.

INTRODUCTION

The importance of PCBs in environmental studies is mainly due to the fact that they have been used in many industrial applications for about forty years, without any precaution to prevent environmental contamination. In addition, their high chemical stability and ability to accumulate in organisms are responsible for long residence times in the environment and for toxic effects on biota, respectively[1–4].

For these reasons a monitoring program of PCBs was begun in 1988 within the Environmental Impact—Chemical Methodologies framework of the Italian Research Programme in Antarctica (PNRA). The main aim was to evaluate the presence of PCBs in Antarctica and to gain a better understanding of the diffusion mechanisms of these contaminants in Antarctica over time by sampling those matrices, such as sediments, which have recorded information on past events in their depth-dependent chemical composition.

The results relevant to the analysis of samples collected during previous italian expeditions in Antarctica (1988–89, 1989–90 and, some of the samples from 1990–91) have already been discussed[5,6].

The present paper deals with the 1990–91 and 1 991–92 expeditions, whose aims were the following:

1990–91 Expedition

- to complete a map on the presence of PCBs in marine sediments in a large area of the Ross Sea (Figures 1 and 2);
- to evaluate the PCB depth profile in marine sediments;
- to confirm the presence of PCBs in Antarctica by sampling soils and lake sediments in the area of Victoria Land around the Italian Base (Figure 2), already studied during previous expeditions;
- to investigate the effect of pack ice melting on the PCB depth profile in coastal sea water at Gerlache Inlet (Figure 3).

1991–92 Expedition

- to verify the effect of pack ice melting on the PCB content in the surface sea water layer at Gerlache Inlet (Figure 3).

The results on the PCB content in marine sediment, lake sediment, soil and sea water samples are presented, including depth profiles in sediments and coastal sea water. An optimized procedure for chromatographic peak assignment is also discussed.

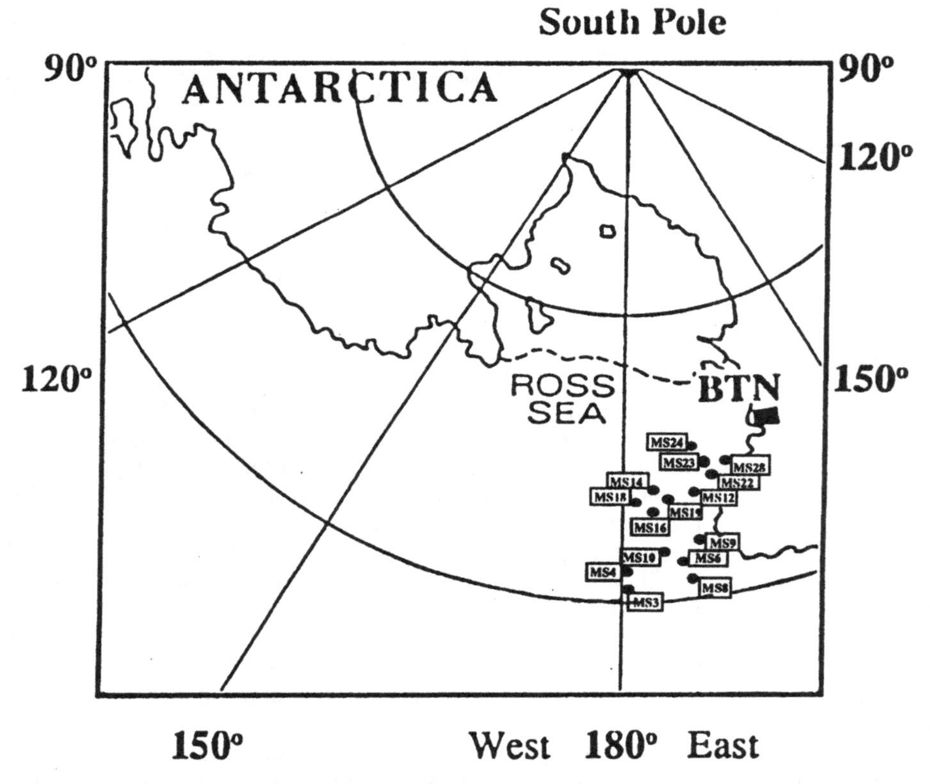

Figure 1 Location of sampling stations of marine sediment (MS) samples gathered during the 1990–91 Italian expeditions in Antarctica. At stations MS10, MS18 and MS22 samples were collected at different depths. (The dotted line represents the pack ice edge).

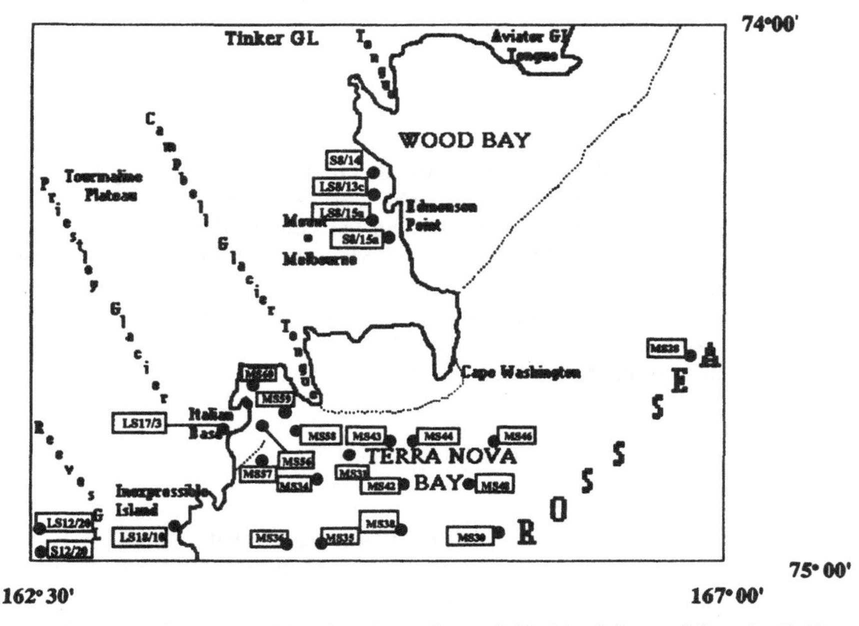

Figure 2 Location of sampling stations of marine sediment (MS), lake sediment (LS) and soil (S) samples gathered during the 1990–91 Italian expeditions in Antarctica. At station MS38 samples were collected at different depths. (The dotted line represents the pack ice edge).

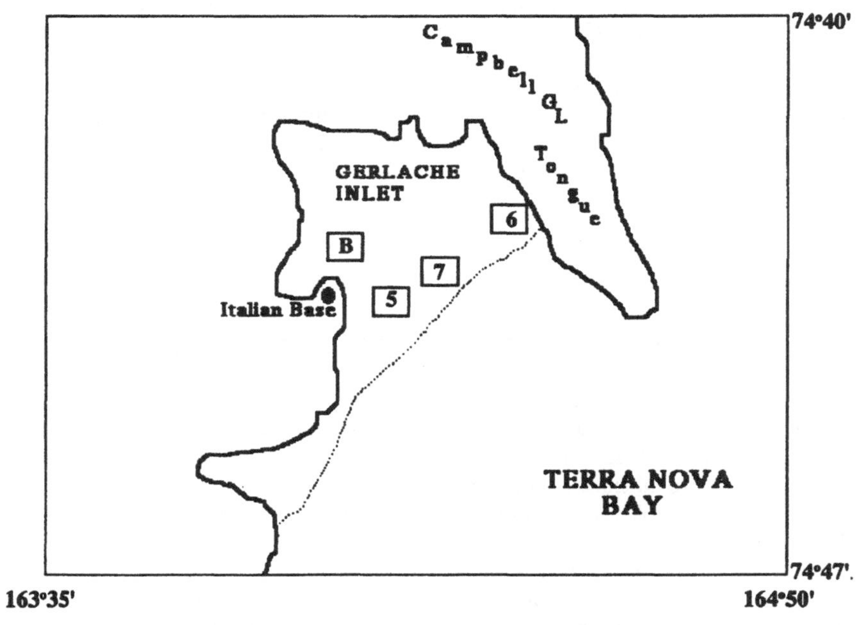

Figure 3 Location of sampling stations of sea water samples gathered before and after pack ice melting during the 1990–91 (Station B) and the 1991–92 (stations B, 5, 6 and 7) Italian expeditions in Antarctica. (The dotted line represents the pack ice edge).

EXPERIMENTAL

Reagents

n-Hexane, acetone and dichloromethane Pesticide Grade; Na_2SO_4 and Hg RPE-ACS; Cu powder RLE and Florisil RS (60–100 mesh) were supplied by Carlo-Erba (Italy). Aroclor 1221, 1232, 1248, 1260 (35 μg/ml) and individual PCB congeners (35 μg/ml) standard solutions were supplied by AccuStandards (USA). Reference marine sediment samples CS-1 and HS-2, containing 1.2 ng/g and 112 ng/g of total PCBs respectively, expressed as Aroclor 1254, were supplied by the National Research Council of Canada. Reference soil samples, containing 91 ng/g of Aroclor 1260, were supplied by Environmental Resources Associated (USA). Reagent pre-treatments are described elsewhere[5].

Apparatus

A supercritical fluid chromatograph (SFC) mod. 3000 (Carlo Erba Strum., Italy), used in the GC mode, and a gas chromatograph (GC) 5160 Mega series (Carlo Erba Strum., Italy), both equipped with automatic cold on-column injection port mod. OC516 and electron capture detector (ECD) were used. Chromatographic separation was always performed on a chemically bonded fused silica capillary column CP-Sil 8CB (Chrompack Italy S.r.l.) 0.25 mm I.D., 0.25 μm film thickness, 50 m length, connected to 2 m long deactivated fused silica capillary pre-column 0.32 mm I.D.. The chromatographic conditions were:column heated at 60°C isothermal for 2 min, then 15°C/min up to 180°C and isothermal for 6 min, 4°C/min up to 220°C and isothermal for 2 min, 5°C/min up to 280°C and isothermal for 25 min; detector temperature 320°C, carrier gas helium, make-up gas nitrogen. A mass spectrometric detector mod. 5971 (Hewlett Packard Italiana, Italy) coupled to a GC was used for the identification and assignment of chromatographic peaks. A Microtrac particle analyzer (Leeds & Northrup Int., USA) equipped with two optical benches, which permits analysis in the 0.12–42 μm and 1.2–300 μm particle size ranges, was used to evaluate the particle size distribution of sediment and soil samples.

Sampling

Sediment and soil samples: Figures 1 and 2 show the sampling stations of marine sediment, lake sediment and soil samples collected during the 1990–91 expedition. Marine sediments were generally collected with a stainless steel grab. In four stations, namely MS10, MS18, MS22 and MS38, samples were collected with a box-corer system and aliquots between 0–15 cm and 15–30 cm were collected, with the exception of station MS22 where aliquots were collected at 0–10, 10–20 and 20–30 cm. Lake sediment and soil samples were collected in the same area already studied in previous expeditions. In particular, lake sediments were collected manually in four small lakes which were generally 30–150 m wide and 60–150 cm deep. Three of the lakes were located along the coast line and one was located 20 Km from the coast (Figure 2); soil samples were also collected manually in three sites near the lakes when possible (Figure 2). All the samples were stored at –20°C in polyethylene containers suitably cleaned and conditioned before use.

Sea water samples: Figure 3 shows the sampling stations of sea water samples collected at different depths (station B: 0.5, 10, 25, and 250 meters) and at the surface (stations B, 5, 6, 7: 0.5 meters deep) during the 1990–91 and 1991–92 expeditions, respectively. Sampling was performed with a "go-flo" system with teflonated bottles or a teflon pumping system. Samples were stored in 20-liter stainless steel containers (Sartorius, mod. SM 17S32) at –20°C.

Analytical methods

The analytical procedures used, including the evaluation of accuracy and precision for PCB determination in the matrices considered, have been discussed elsewhere[5,6]. In particular, sea water samples were extracted with n-hexane, and the stainless steel container also rinsed with the same solvent. These aliquots were mixed together, dried, and reduced to a volume of about 1.5 ml. As far as sediments and soils is concerned, 20 g of sample were weighed, made homogeneous and divided into two aliquots of 15 g and 5 g each. The 5 g aliquot was used to evaluate the percentage of water in the sample and to perform particle size analysis. The 15 g aliquot was extracted with 1:1 n-hexane/acetone mixture in an ultrasonic bath. The extract was treated with Cu powder and Hg for sulphur removal, and reduced to about 1.5 ml.

For all the samples, the final extract was loaded on a Florisil column, from which PCBs were selectively eluted with 10 ml of n-hexane. The eluate was concentrated at 100 μl right before the analysis performed by GC-ECD or GC-MS.

RESULTS AND DISCUSSION

Identification of PCBs

The content of individual PCB congeners in Antarctic sediment, soil and sea water samples is very low, and is generally below the detection limit of GC-MS. This occurs even if large quantities of samples are extracted (20 g for sediment and soil samples, and 20 l for sea water samples) and the final extracts are concentrated at a final volume as low as 100 μl which can still be weighed with an acceptable accuracy. This makes GC-ECD the only useful technique for quantitative analysis. In this case peak assignement is made on the basis of the expected retention times within a fixed time window. A time window of 0.1 min was used in previous works as generally reported in literature[1]. If we consider the chromatogram B shown in Figure 4 it becomes evident that at this PCB concentration level there could be a lot of interfering signals which may affect peak assignment.

In order to minimize incorrect peak assignment it is therefore extremely important to calculate a statistically significant time window. This was done by calculating the mean retention times (RT_{mean}) referred to two internal standards, namely PCB36 and PCB209, and the sample standard deviations (s_{RT}) of nine selected PCB congeners on the basis of the results relating to 5 injections of standard solutions (Table 1). The corresponding 95% confidence limits (±0.012 min) were then obtained and used as a time window for peak assignement.

The correctness of this assumption was confirmed by analyzing standard solutions and certified sediment samples: the relevant chromatograms were evaluated by using both the time windows 0.1 and 0.012 min, and two values of concentrations were calculated.

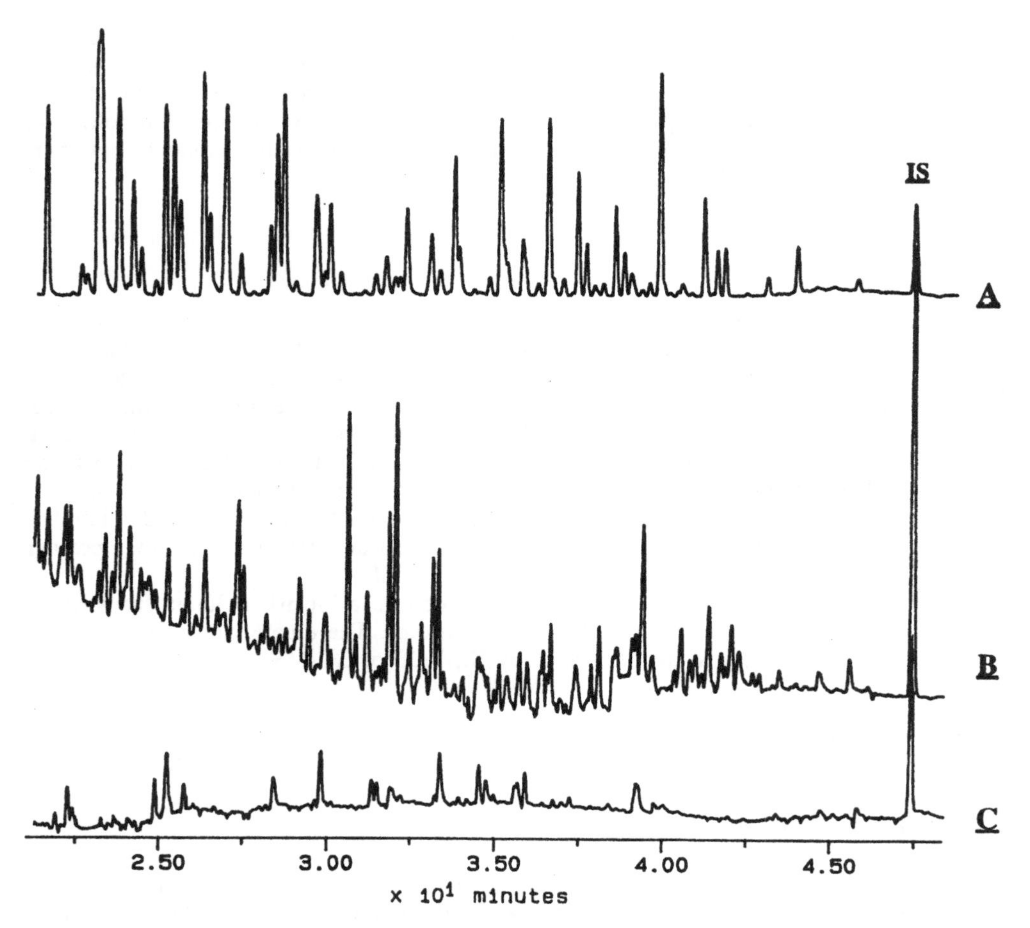

Figure 4 ECD-gas chromatograms of *A*) Aroclor mixture standard solution (Aroclor) 1221, 1232, 1248 and 1260: about 300 pg of total PCBs injected); *B*) extract of an Antarctic sea water sample (30 pg estimated total PCBs injected); *C*) blank. (*IS* = internal standard; for experimental conditions see text).

These two values were both in agreement with the expected ones and their differences were about 5% and 10% for standard solutions and certified sediment samples, respectively. The same procedure was used to evaluate the chromatograms of all the samples collected, which showed much higher differences. In particular, the PCB concentration of sediment and sea water samples decreased typically by a factor of 1.3 and 3 respectively, when the time window was decreased from 0.1 to 0.012 min.

At present, a supercritical fluid extractor (SFE) coupled with a cold trap-GC-MS system is being developed in our laboratory which will permit us to inject higher volumes (50–100 µl) into the chromatographic column, thus allowing the determination of organic micropollutants in environmental samples by a mass spectrometric detector even at low pg/l levels.

Table 1 Reproducibility of the retention times of nine PCB congeners. The mean values (RT_{mean}) and the sample standard deviations (S_{RT}) were obtained on 5 repeated measurements.

PCB (IUPAC number)	*RT_{mean} (min)*	*S_{RT} (min)*
28	27.053	0.011
52	29.047	0.010
101	34.957	0.012
118	39.794	0.011
153	41.643	0.009
105	41.934	0.008
138	43.935	0.009
156	48.154	0.009
180	49.270	0.011

Analysis of sediment and soil samples

Tables 2 and 3 show the results relevant to marine sediments, and Table 4 those relevant to lake sediments and soils. Actually, the amount of PCBs present in samples such as sediments or soils, is much more likely to be related to the particle surface area per volume unit, where they are adsorbed, than to the mass unit[6]. For this reason, the concentration of each sample, expressed in pg/g dry weight, was normalized by dividing it for the relevant calculated specific surface area (CS) as obtained by particle size analysis[5,6]. The following concentration ranges—expressed in (pg/g)(m^2/cm^3) dry weight—were found: marine sediments 90–230 (mean value: 145), lake sediments 170–240 (mean value: 218) and soils 120–160 (mean value: 137). In the four stations where marine sediments were collected at different depths, a concentration of about 100–200 (pg/)(m^2/cm^3) was observed in a surface layer of about 10–15 cm, while in deeper layers PCBs were below the detection limit. Only for sample MS38 a very low quantity of PCBs—about 20 (pg/g)(m^2/cm^3)—was measured in the second layer.

From all these results the following conclusion can be drawn:

- no significant differences of normalized PCB content in marine sediments among stations located in open sea [stations from MS3 to MS28, mean value: 142 (pg/g)(m^2/cm^3)] and those located closer to the coastal line [stations from MS30 to MS60, mean value: 148 (pg/g)(m^2/cm^3)] were observed. This slight and constant contamination may exclude any direct source of PCB pollution in Antarctica. The same situation was also observed for lake sediments and soils;
- the results relevant to depth profiles in marine sediments showed that PCBs were confined in a surface layer of about 10 cm. This result is supported by the fact that PCBs began to be used at an industrial level in 1930 and a sediment layer 10-cm deep corresponds to about 100 years according to the sedimentation rate of the area under study which was 0.05–0.1 cm/year as estimated by using the ^{210}Pb method[7]. The very low content found in the second layer of station MS38 may be explained by considering bioturbation processes; although a slight contamination of this sample cannot be excluded “a priori”;
- lake sediments show the highest normalized PCB content. These results, as already stated[6], might be explained by considering the nature of Antarctic lakes, which are

formed during the deglacial season, and taking into account the PCBs trapped in the ice matrix during its formation which come from the atmospheric particulate[8–10].

Analysis of sea water samples

The results relevant to the samples collected during the 1990–91 expedition before and after pack ice melting at different depths, namely 0.5, 10, 25 and 250 m, have already been described in a previous paper[6]. However all the chromatograms were reprocessed according to the procedure for peak assignment described above, which showed much lower PCB content (Table 5). These results highlighted the low PCB contamination of the area under observation (typically 150 pg/l). In particular, the PCB content along the water column before pack ice melting ranged in a narrow interval around 140 pg/l, with a consistent increase near the sea bed at 250 m, which is probably due to remobilization

Table 2 Total PCB content of marine sediment samples gathered at Terra Nova Bay and Ross Sea during the 1990–91 Italian expedition in Antarctica (the relative standard deviation of mean values is reported in brackets).

Sampling station	*PCBs (pg/g dry weight)*	*CS (m^2/cm^3)*	*PCBs/CS (pg/g)/(m^2/cm^3)*
MS3	100	0.80	120
MS4	140	0.71	200
MS6	110	0.78	140
MS8	80	0.76	110
MS9	90	0.68	130
MS10	60	0.61	100
MS12	150	0.66	230
MS14	50	0.50	100
MS16	90	0.45	200
MS18	50	0.35	140
MS19	60	0.51	120
MS22	50	0.50	100
MS23	70	0.48	150
MS24	100	0.56	180
MS28	60	0.53	110
MS30	160	0.44	170
MS31	60	0.37	160
MS34	30	0.23	130
MS35	40	0.42	100
MS36	30	0.27	110
MS38	110	0.56	200
MS42	70	0.37	190
MS43	40	0.39	100
MS44	50	0.54	90
MS46	80	0.52	150
MS48	70	0.49	140
MS56	70	0.33	210
MS57	60	0.35	170
MS58	30	0.28	110
MS59	60	0.50	120
MS60	100	0.45	220
mean value	75 (45%)		145 (28%)

Table 3 Depth profile of total PCB content in marine sediment samples gathered at Terra Nova Bay and Ross Sea during the 1990–91 Italian expedition in Antarctica.

Sampling station	*Depth (cm)*	*PCBs (pg/g dry weight)*	*CS (m^2/cm^3)*	*PCBs/CS (pg/g)/(m^2/cm^3)*
MS10	0–12	60	0.61	100
	13–25	n.d.	0.89	n.d.
MS18	0–14	50	0.35	140
	15–28	n.d.	0.61	n.d.
MS22	0–10	50	0.50	100
	11–20	n.d.	0.58	n.d.
	21–30	n.d.	0.53	n.d.
MS38	0–15	110	0.56	200
	16–34	15*	0.75	20*

n.d. = not detectable; (*) approximate value

Table 4 Total PCB content of lake sediment and soil samples gathered at Terra Nova Bay and Victoria Land during the 1990–91 Italian expedition in Antarctica (the relative standard deviation of mean values is reported in brackets).

Sampling station	*PCBs (pg/g dry weight)*	*CS (m^2/cm^3)*	*PCBs/CS (pg/g)(m^2/cm^3)*
lake sediment			
LS8/13c	60	0.35	170
LS8/15a	90	0.39	230
LS17/3	120	0.43	280
LS18/10	70	0.40	170
LS12/20	90	0.38	240
mean value	86(27%)		218 (22%)
soil			
S8/14	50	0.32	160
S8/15a	40	0.33	120
S12/20	70	0.52	130
S17/3	60	0.45	130
mean value	55 (23%)		135 (13%)

Table 5 Total concentration of PCBs in sea water samples gathered at Terra Nova Bay – Gerlache Inlet before and after pack ice melting during the 1990–91 Italian expedition in Antarctica.

Sampling station	*Depth (m)*	*PCBs (pg/l)*	
		Before pack ice melting	*After pack ice melting*
B	0.5	130	170
B	10	150	120
B	25	140	160
B	250	210	220

Table 6 Total concentration of PCBs in surface sea water samples gathered at Terra Nova Bay – Gerlache Inlet before and after pack ice melting during the 1991–92 Italian expedition in Antarctica.

Sampling station	*PCBs (pg/l)*	
	Before pack ice melting	*After pack ice melting*
5	150	290
6	220	250
7	170	230
B	140	370
mean value	170 (21%)	285 (22%)

processes involving sediments. In this case, no significant change in PCB content was observed after pack ice melting, while during the 1991–92 expedition an increase of about 70% in the PCB content in the surface layer of sea water after pack ice melting was found (Table 6). The pack ice and sea water samples collected during the 1993–94 expedition are currently being analyzed and this will confirm whether this increase can be explained once again by considering the transfer to sea water of PCBs associated with the atmospheric particulate trapped in the pack ice.

Acknowledgements

We wish to express our sincere appreciation to the researchers of the "Environmental Impact–Chemical Methodologies" team for their accurate and skillful work, and to the technical staff of the 1990–91 and 1991–92 Italian expeditions in Antarctica for their indispensable help during sample collection. This work was financially supported by ENEA (Rome) under the Italian Research Programme in Antarctica.

References

1. M. D. Erickson, *Analytical Chemistry of PCBs*, (Butterworth Publishers Stonehouse, MA, USA, 1986), pp. 24.
2. J. Albaiges (editor), *Environmental Analytical Chemistry of PCBs*, (Gordon and Breach, Reading, UK, 1993), p. 409.
3. J. T. Borlakoglu and R. R. Dils, *Chem. Brit.*, 815–818 (Sept. 1991).
4. D. E. Tillitt, J. P. Giesy and G. T. Ankley, *Environ. Sci. Technol.*, **25**, 87–92 (1991).
5. R. Fuoco, M. P. Colombini and C. Abete, *Ann. Chim. (Rome)*, **81**, 383–394 (1991).
6. R. Fuoco, M. P. Colombini and C. Abete, *Intern. J. Environ. Anal. Chem.*, **55**, 15–25 (1994).
7. M. Ravaioli, *Private Communication*, (1994).
8. J. C. Duinker and F. Bouchertall, *Environ. Sci. Technol.*, **23**, 57–62 (1989).
9. S. W. Fowler, *PCBs and the Environment: the Mediterranean marine Ecosystem*, in *PCBs and the Environment* (J. S. Waid, ed., CRC Press, Boca Baton, Fl, USA, Vol III, 1986), pp. 209–239.
10. P. Larsson and A. Sodergren, *Water, Air and Soil Pollution*, **36**, 33–37 (1987).

HALOCARBONS IN ANTARCTIC SURFACE WATERS AND SNOW

L. ZOCCOLILLO, L. AMENDOLA and G. A. TARALLO

Dipartimento di Chimica, Università "La Sapienza", P. le Aldo Moro 5, 00185 Rome, Italy

Surface and pit snow samples, collected from Terranova Bay area in Antarctica during the Italian expeditions of 1991–92 and 1992–93, were analysed, for halocarbons, namely tetrachloromethane, trichloroethylene and tetrachloroethylene. The results obtained (including those related with lake and ice water samples previously reported) were evaluated with respect to the worldwide distribution of these compounds and their diffusion to a global scale. Important innovations concerning sensitivity and reproducibility of the analytical method, are also reported.

KEY WORDS: Antarctica, natural water, snow, halocarbons.

INTRODUCTION

In a previous work[1], traces of tetrachloromethane, trichloroethylene and tetrachloroethylene were found in lake and ice water samples collected in the Terranova Bay area in Antarctica during the Italian expeditions in 1988–89, 1989–90, 1990–91. The above compounds were selected as pollution indicators in remote areas because large amounts have been released into the environment over the past fifty years, and their volatility may contribute to their diffusion to a global scale. The possibility of sampling these chemicals at very low concentrations (ng/L) in natural water samples makes also attractive their determination. The present paper reports on the analysis of surface and pit snow samples collected during the Italian Antarctic expeditions of 1991–92 and 1992–93. Research on these compounds was extended to snow because this is the only compartment in which pollution is deposited on the Antarctic continent. The work also aimed at improving analytical methods, as regards sensitivity, reproducibility and sample volumes used.

EXPERIMENTAL

Sampling

Two surface snow and one ice samples were taken during the brief Italian expedition to Antarctica in 1991–92. In the 1992–93 expedition, seven snow samples were taken from a three metre pit, one sample every 0.5 m (Table 1). Stainless steel cylinders (vol = 10 L

Table 1 Samples and sampling site for Antarctic ice and snow.

Expedition	*Sample*	*Station*	*Altitude*	*Lat. S*	*Long. E*
1991–1992	Ice	Reeves Glacier	—	74° 43'	162° 45'
1991–1992	Surface snow	Priestley Glacier	—	74° 08'	162° 48'
1992–1993	Snow pit	Hercules Nevè	2960 m	74° 06'	165° 28'

and i.d. = 20 cm) were used for the snow samples (Figure1). The container lids had an internal diameter equal to the outer diameter of the containers. A silica rubber seal allowed air-tight closure of the cylinders. The shape of the cylinder allowed direct snow sampling from the pit sides. The quantity of snow taken for each container corresponded to approx. 2 L of water. The same type of container was used for surface ice, which was sampled using an ice-pick. All samples were kept at –20°C before analysis.

Apparatus and materials

A Hewlett-Packard 5890 series II gas chromatograph was used for sample analysis. The gas chromatograph was equipped with a HP-5 capillary column (Crosslinked 5% PhMe Silica), 25 m × 0.32 mm × 0.52 μm film thickness, an on-column injector and an ECD detector. A Hewlett-Packard GC-MS 5989 system with Chemstation HP59940A was used to confirm the peak identities. n-Hexane for organic residue analysis (J. T. Baker Chemical Co.) was used for extraction, since it was free of volatile halocarbons. Tetrachloromethane, trichloroethylene and tetrachloroethylene for analysis (MERCK) were used for the standard solutions of halocarbons in n-hexane. All the glassware was washed in chromic mixture and distilled water, and the water residues were removed under hot air.

Extraction

Extraction was carried out on the melted snow and ice samples using a round flask (0.5 L) in which 0.5 L of water were extracted with 0.5 mL of hexane (Figure 2a). After 15 min of vigorous shaking using a magnetic stirrer and after separation of the phases, the organic extract was recovered using the device shown in Figure 2b. The microextractor has two parallel tubes of different diameters in the upper section. Tube A, in which the organic extract separation actually takes place, has an internal diameter of 0.5 cm and a total length of 20 cm. Tube B, with an internal diameter of 0.8 cm, is wider in the upper portion to facilitate the introduction of water to the flask. Tube B is 5 cm longer than tube A in the lower portion, so that when the microextractor replaces the flask stopper, the lower portion of tube B lies below the hexane microlayer.

The water introduced in the upper part of tube B ends up below the hexane microlayer and practically does not come into contact with it. The addition of water, free of the test substances, causes the liquid to rise until the hexane microlayer is channelled into the upper part of tube A. Alternatively, another microextractor version has been made

Figure 1 Stainless steel container used for ice and snow sampling (Inox Sabat, Bologna, Italy).

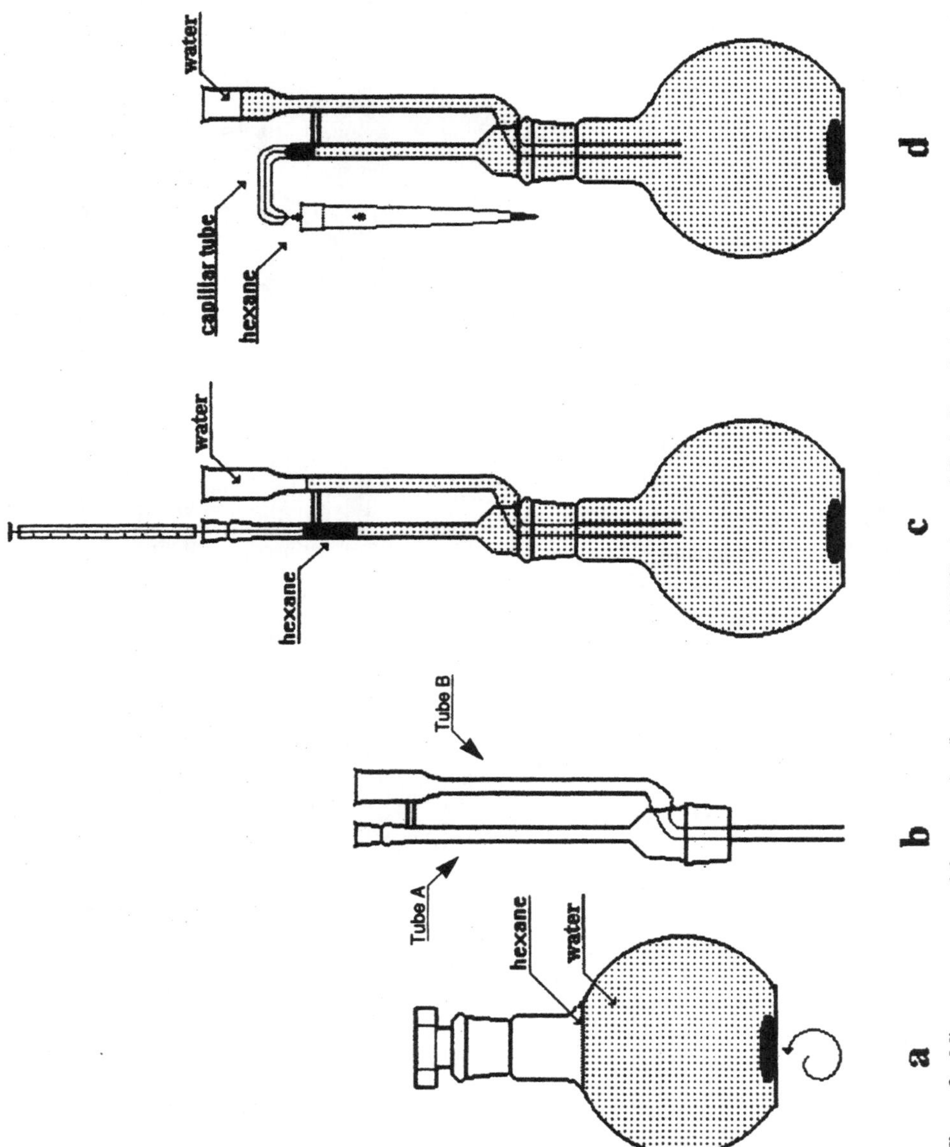

Figure 2 Microextractors used for extraction (hexane/water 1/1000); a) round flask; b) typical microextractor; c) extract sampling for direct injection in the gas chromatograph; d) extract separation for possible further extract concentration.

which allows complete recovery of the organic extract and avoids sampling using a Pasteur pipette (Figure 2d). In the new version, tube A is 5 cm shorter and ends with a U-shaped capillary. Because of the higher surface tension of water compared with hexane, when the meniscus of the water lying beneath the hexane reaches the top of the capillary the flow of liquid stops.

Without concentrating the extract, the extraction system shown in Figure 2c allows obtaining the required halocarbons in hexane at the µg/L level (measurable with GC-ECD and GC-MS) with respect to the ng/L level present in the original water sample. The microextractor in Figure 2d is useful for separating all the extract for further research of other less volatile compounds.

GC-ECD-MS analysis

Analysis of the snow and ice organic extracts was carried out using a GC-ECD apparatus (Carrier gas: N_2; temperature: 40°C for 8 min, 10°C/min to 120°C, and hold for 1 min). The detection limit of the ECD detector used for the investigated species is about 0.1 pg. At least a concentration of 0.1 ng/l of tetrachloromethane, trichloroethylene and tetrachloroethylene in water is detectable using the described extraction system. The standard deviation of the halocarbons measured concentration is about 10%. A GC-MS apparatus under the EI-SIM mode, was used to confirm the peak identities (Carrier gas: He; temperature: same conditions of GC-ECD analysis). In order to allow the use of a high resolution column (0.20 mm i.d. with the on-column injector) a section of fused silica column, without fixed phase was mounted in the injector. This section had a length of 10 cm and an internal diameter of 0.53 mm, and was connected to the analytical column (0.20 mm i.d.) via a pressfit. The 0.53 mm column section acted as a microinjector suitable for conventional 0.44 mm diameter needles. Figure 3B shows a typical GC-MS in SIM chromatogram of an extract of Antarctic snow.

RESULTS AND DISCUSSION

Table 2 reports the results of the analyses carried out on the snow and ice samples from the Italian expeditions of 1991–92 and 1992–93. The results confirm the conclusions reported in a previous work[1], i.e. that the Terranova Bay area is widely polluted by tetrachloromethane, trichloroethylene and tetrachloroethylene. The presence of these halocarbons in Antarctica is a further proof of their ubiquitous occurrence in the environment and, owing to their origin in inhabited areas of the world, their diffusion at a global scale. Although a lot of natural waters (such as spring waters, tap waters and mineral waters) were analysed, no one of the three investigated species was found above the detection limit of the method.

Diffusion model at global level

The environmental behaviour of the substances analysed is well-known in temperate regions and is related to their chemical and physical properties (Table 3). Once released into the environment, the high vapor pressure of the three compounds determines their preferential distribution: in air and with only low percentages in water. Therefore, the main way they are transported over great distances is via the atmosphere. Mackay and

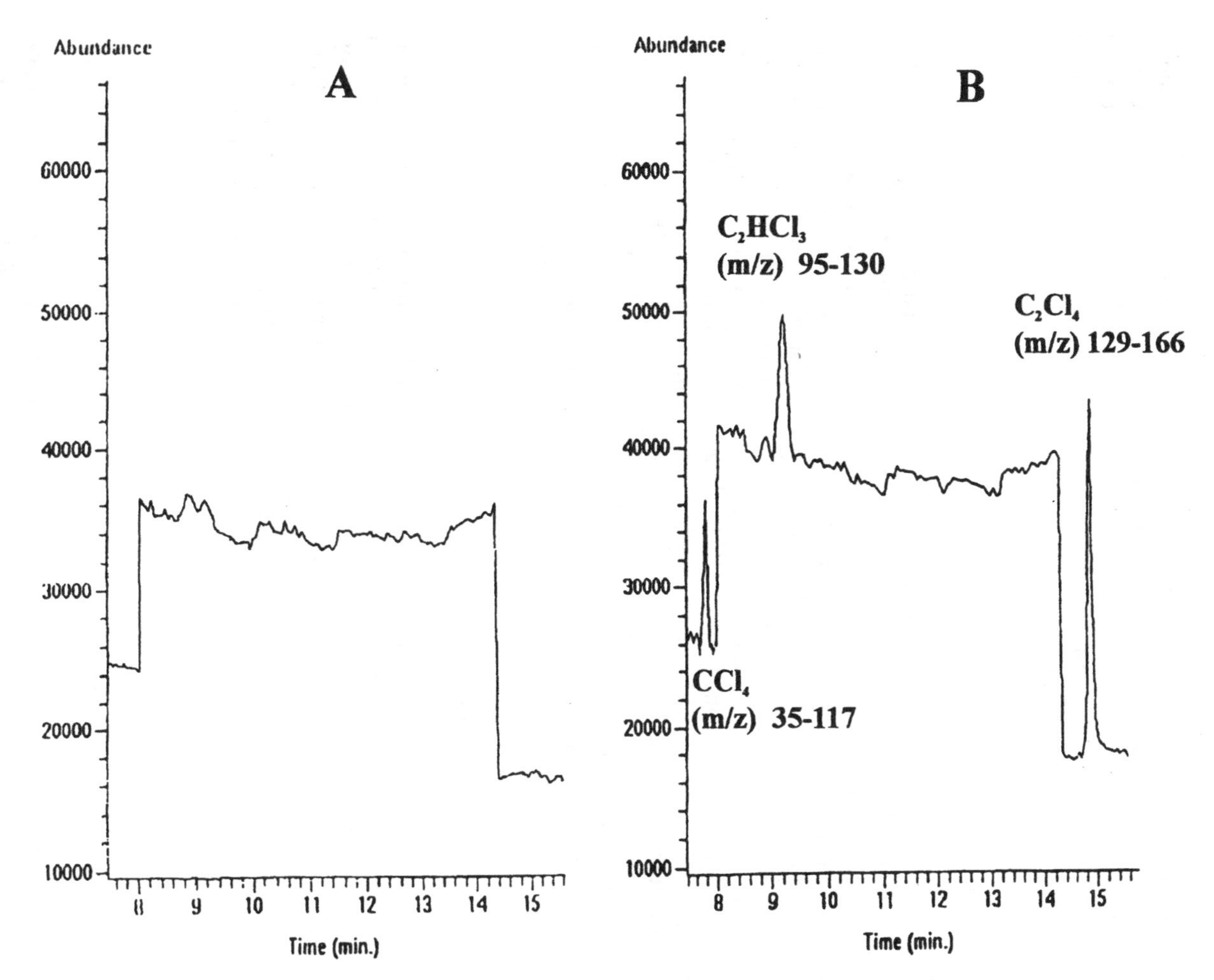

Figure 3 SIM-GC-MS chromatogram: A) n-hexane used for extraction; B) typical Antarctic snow extract.

Table 2 Halocarbons in antarctic ice and snow samples (ng/L).

Compound (Expedition)	*Tetrachloromethane*		*Trichloroethylene*		*Tetrachloroethylene*	
	'91–'92	*'92–'93*	*'91–'92*	*'92–'93*	*'91–'92*	*'92–'93*
Surface snow (Priestley Glacier)	4.6	–	14	–	15	–
Snow (depth 0.5 m) (Priestley Glacier)	2.5	–	8.5	–	5.0	–
Surface ice (Reeves Glacier)	3.3	–	11	–	7.5	–
Snow pit (depth 0.0 m) (Hercules Nevè)	–	12	–	5.7	–	7.4
Snow pit (depth 0.5 m) —	–	7.9	–	2.9	–	2.5
Snow pit (depth 1.0 m) —	–	8.8	–	8.3	–	5.7
Snow pit (depth 1.5 m) —	–	12	–	8.9	–	7.2
Snow pit (depth 2.0 m) —	–	6.3	–	3.9	–	3.7
Snow pit (depth 2.5 m) —	–	14	–	8.0	–	3.6
Snow pit (depth 3.0 m) —	–	9.8	–	9.0	–	6.6

Table 3 Selected chemical-physic properties of tetrachloromethane, trichloroethylene and tetrachloroethylene.

Compound/Property	CCl_4	$CHCl = CCl_2$	$CCl_2 = CCl_2$
Vapor pressure	11.94 kPa (20°C) (4)	5.8 kPa (20°C) (5)	1.3 kPa (14°C) (4)
Solubility in H_2O	0.8 g/L (25°C) (4)	0.2 g/L (20°C) (4)	0.2 g/L (25°C) (4)
k_{OH}	$<10^{-15}$ cm^3s^{-1} (5)	$2.2*10^{-12}$ cm^3s^{-1} (6)	$1.7*10^{-13}$ cm^3s^{-1} (6)
Lifetime	47 y (6)	0.021 y (6)	0.43 y (6)

Paterson[2,3] have established a behavioural model for temperate regions applicable to trichloroethylene. This model shows that, once released into the atmosphere, 92% of the trichloroethylene is carried in the air and only 8% in water. 54% of the trichloroethylene present in the air is then eliminated by OH radicals and UV radiation while the remaining 46% is carried by air currents to nearby regions. Because tetrachloromethane and tetrachloroethylene have similar chemical and physical properties to trichloroethylene, their environmental behaviour will be similar except for their lifetimes in the troposphere (Table 3), according to their reactivity with the OH radical (k_{OH}) (Table 3).

A chemical compound may be carried by the atmosphere as a vapor or through an atmospheric particulate matter in which the substance is adsorbed. The adsorbed percentage of the chemical substance in the particulate matter is mainly determined by its vapor pressure. Since chemical substances with vapor pressures greater than 1.3 10^{-2} kPa, are completely in the air phase[7] and considering the specific values of the three compounds studied (Table 3), the atmosphere can only carry the substances in the vapor state.

The phenomena regulating deposition of pollutants from air to snow are extremely complex and may be divided into two groups: wet and dry deposition[8–10]. In the former case, there is an equilibrium state between air and water phases on the water microlayers present, in summertime, on surface snow crystals or on atmospheric particulate matter. Dry deposition, on the other hand, occurs when snow forms, a very rare phenomenon in Antarctica, and via direct deposition onto the snow surface. In the latter case, the main factors which determine the deposition rate are: the surface area exposed, roughness of surfaces, and wind speed above these surfaces. Since Antarctica has a vast uneven, air permeable, snow surface and has winds that may exceed 100 km/h, dry deposition on the snowpack is certainly an important phenomenon in the accumulation of pollutants. Because the three halocarbons studied are present in the troposphere as vapors, it is likely that the main way they are deposited in snow is by dry deposition. The presence of the three halocarbons in pitsnow of up to 3 m deep at uniform concentration levels, within the experimental errors (Table 2) would confirm the predominance of dry deposition on snow surfaces. Up to a depth of three metres, permeability of snow to air is practically the same in Antarctica.

The halocarbon concentrations found in the water samples are similar and, in some cases, even greater than those found for the three substances in water samples taken from areas in Italy that are very remote from sources of pollution[1]. This may be explained by the very efficient deposition and accumulation mechanism in Antarctica that is favoured by the specific environmental characteristics, despite the fact that atmospheric concentrations of the three halocarbons are much lower than for inhabited areas of the world.

Acknowledgement

This work was performed in the framework of "Environmental Contamination Project-National Research Program in Antarctica", Italy.

References

1. L. Zoccolillo and M. Rellori, *Intern. J. Environ. Anal. Chem.*, **55**, 27–32 (1994).
2. D. Mackay, S. Paterson, B. Cheung and W. B. Neely, *Chemosphere*, **14**, 335–374 (1985).
3. D. Mackay and S. Paterson, *Environ. Sci. Technol.*, **25**, 427–436 (1991).
4. *Kirk-Othmer Encycl. Chem. Technol.* (Wiley-Interscience Pub. New York USA, 1980), 3rd Ed.
5. *Ulmann's Encycl. Ind. Chem.* (VCH Weinheim Germany, 1985), 5th Ed.
6. World Metereological Organization, Global Ozone Research and Monitoring Project, Report N° 25; Scientific Assessment of Ozone Depletion: 1991 (1991).
7. T. F. Bidleman, *Environ. Sci. Technol.*, **22**, 361–366 (1988).
8. D. J. Gregor, *Nato ASI Ser.*, Ser. G **28**, 323–357 (1991).
9. S. H. Cadle, *Nato ASI Ser.*, Ser. G **28**, 21–66 (1991).
10. L. A. Barrie, *Nato ASI Ser.*, Ser. G **28**, 1–20 (1991).

OCCURRENCE OF OXYGENATED VOLATILE ORGANIC COMPOUNDS (VOC) IN ANTARCTICA

P. CICCIOLI*, A. CECINATO, E. BRANCALEONI, M. FRATTONI

Istituto sull'Inquinamento Atmosferico del C.N.R., Area della Ricerca di Roma, Via Salaria Km 29.300, C. P. 10, 00016 Monterotondo Scalo, Italy

F. BRUNER and M. MAIONE

Università degli Studi di Urbino- Centro di Studi per la Chimica dell'Ambiente e le Tecnologie Strumentali Avanzate c/o Istituto di Scienze Chimiche, Piazza Rinascimento, 6. 61029, Urbino, Italy

Polar and non-polar VOC present in six different sites located near Terra Nova Bay in Antarctica were determined by HRGC-MS. 76 different components were positively identified. Among them, particularly important are oxygenated components (free acids, aldehydes, ketones, alcohols, esters, ethers and furanes) as they account for the largest portion of the organic fraction. Biogenic emission seem to be the major source for many of them. This would explain their ubiquitous occurrence in the troposphere.

KEY WORDS: Oxygenated VOC, Antarctica, biogenic emission, ubiquitous air components.

INTRODUCTION

The ubiquitous occurrence of semi-volatile carbonyl compounds in the atmosphere has been recently proposed[1] on the basis of field investigations carried out in urban[1,2], rural[1], forest[1,2] and remote areas[2,3]. Levels in air have been attributed most to biogenic emission from vegetation[1]. Results of research carried out in the Arctic region (Spitzbergen Islands)[4] and Nepal (Himalaya)[4,5] have also shown that, in addition to carbonyl compounds, many other classes of oxygenated VOC are present in remote environments. More than 33 different components including free acids, alcohols, metoxy-alcohols, esters, ethers and furans were positively identified[4-6] in these samples. Since many of them have been also found in plant[7-10] and ocean emissions[11], release from biogenic sources provides a reasonable explanation for their occurrence in remote areas. Due the high levels reached by oxygenated VOC in air, it is possible that VOC emissions from biogenic sources are much larger than what is presently believed and it can strongly contribute to tropospheric ozone formation[6]. Since Antarctica does not possess significant man-made sources of its own and its long distance from polluted areas largely prevent transport of compounds characterized by short-lifetime in the atmosphere, it represents the ideal site for investigating whether biogenic emissions are truly responsible for the occurrence of oxygenated VOC in the troposphere. In this paper, the

VOC composition found in 6 different sites located near the Italian Base installed at Terra Nova Bay will be reported. Data on oxygenated VOC will be discussed and compared with the results obtained in other remote sites.

EXPERIMENTAL

Site description

All samples were collected in the Ross Sea region (Northern Victoria Land) in an area ranging between 74 to 75°S and 162 to 165°E. Six sites were chosen and they are listed in Table 1. Distance and direction from the Italian Base and elevation from the sea level were the main criteria used for their selection. A map showing their location is reported in Figure 1. In Table 1 the exact position of sites, data of sampling and atmospheric conditions in which collection occurred are reported.

Sampling and analysis of VOC

Air samples were collected in stainless steel canisters (BRC, Rasmussen, Hillboro, OR, USA) whose internal surfaces were electro-polished to provide a passivated chromium-nickel oxide layer chemically inert toward polar VOC. Air was transferred into the canisters by using an ultra-clean pump (model FC1121, BRC, Rasmussen, Hillboro, OR, USA) whose contact surfaces were also made in stainless steel. The various parts of the sampling apparatus were assembled according to the scheme reported in Figure 2a. Nupro valves (Nupro Co., Willoughby, OH, USA) were used for connecting canisters with the pump and outlet air and for preventing their contamination from outdoor and indoor sources before and after sampling. Canisters having volumes of 0.85 and 16 L were used in this campaign. They were washed three times with external air before samples were collected at 3 atm. After sampling, canisters were tightly closed, stored in special boxes at room temperature and sent to the laboratory for the analysis.

Aliquots of the total sample (1–2 L) were enriched on traps filled with solid sorbents to meet the sensitivity of the analytical system. They were connected to the canisters according to the scheme reported in Figure 2b. A needle valve was used to keep the flow

Table 1 Sampling sites of VOC at Terra Nova Bay and experimental conditions in which collection was performed.

No.	*Station*	*Lat.* *°S*	*Long.* *°E*	*Altitude* *m*	*Day*	*Volume* *L*	*T* *°C*	*W.S.* *m/s*
1	Tourmaline Pl.	74°08'	163°26'	1650	11.21.93	16	–10	< 1
2	Kay Island	74°04'	165°19'	s.l.	11.22.93	0.85	–10	< 1
3	Edmonson Point	74°20'	165°07'	s.l.	10.27.93	0.85	–6	< 1
4	Mount Melbourne	74°20'	163°20'	1130	11.04.93	0.85	–20	< 1
5	Starr Nunatak	75°54'	162°33'	96	11.18.93	0.85	0	< 1
6	Cape Russel	74°55'	163°50'	70	11.25.93	0.85	0	> 8 SW

All sites were located 30 to 150 km away from the Italian Station at Terra Nova Bay to reduce the influence of man-made sources.

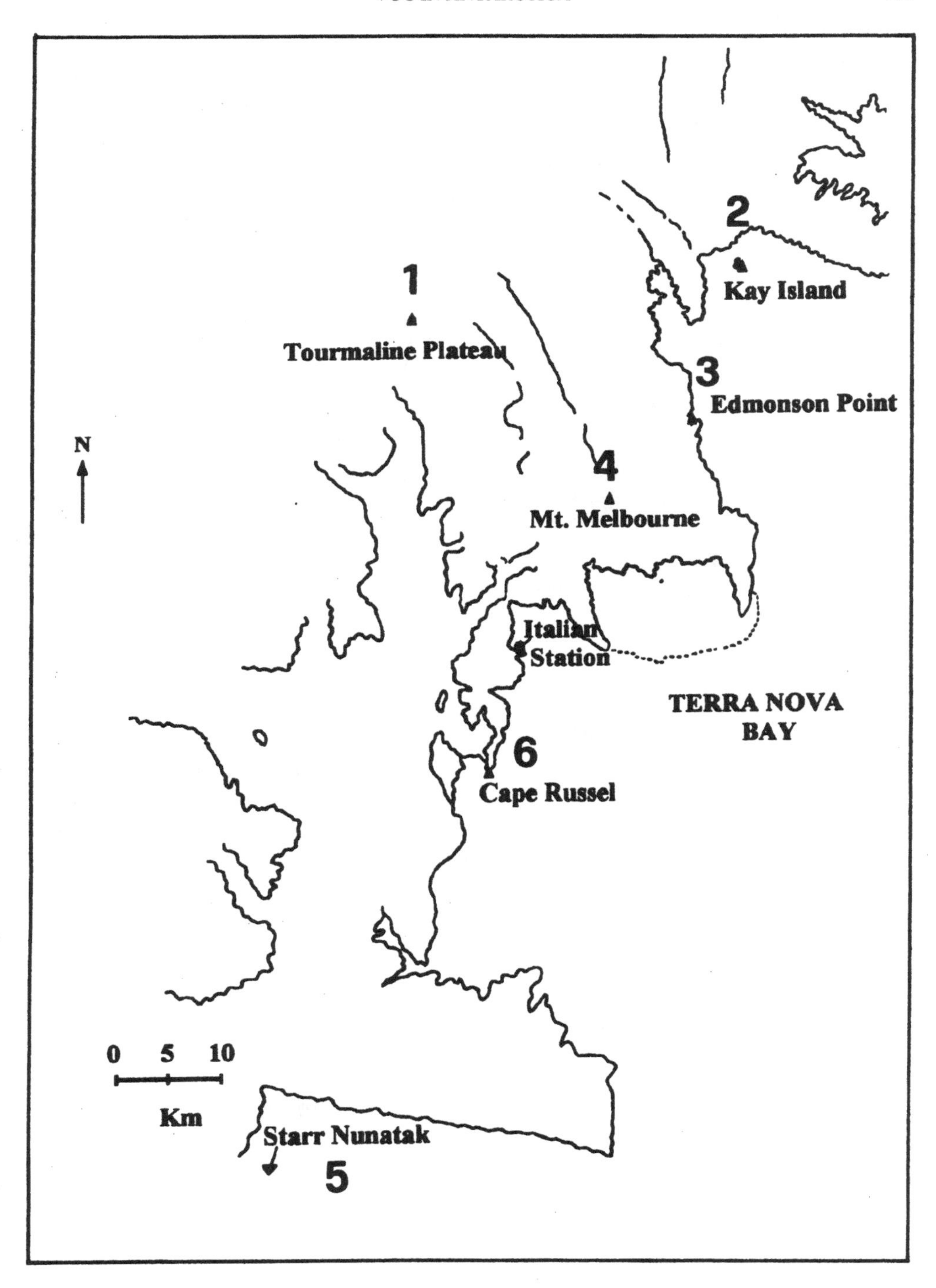

Figure 1 Map showing the location of the VOC sampling sites at Terra Nova Bay.

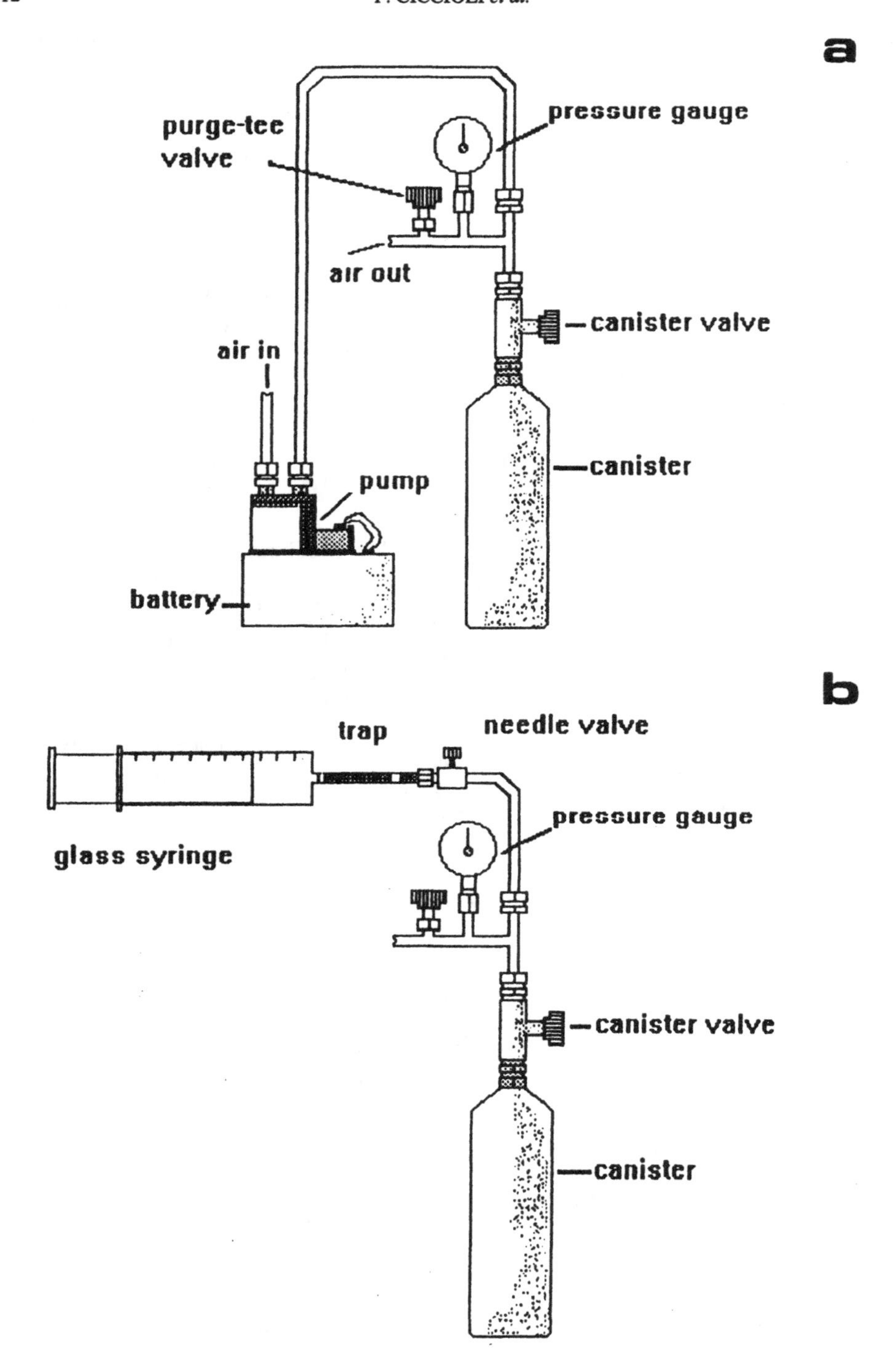

Figure 2 Diagram showing the collection of VOC on canisters (a) and procedure used to transfer them into adsorption traps (b).

rate constant through the trap at a rate of ca. 100 mL/min. The volume sampled was measured with a 250 mL glass syringe connected to the trap outlet. Cartridges were made by glass tubes filled with Carbotrap C (0.034 g) and Carbotrap (0.17 g) supplied by Supelco (Supelco Co, Bellefonte PA, USA). Particles ranging between 20 and 40 mesh were used.

VOC were injected into the GC unit by means of a Chrompack (Chrompack, Middleburg, The Netherlands) purge and trap injector adapted for air analysis. The desorption unit was connected to a Hewlett Packard gas chromatograph (Hewlett Packard, Palo Alto, CA, USA) model 5890 using a 5970 quadrupole mass spectrometer as detection system. GC separations were carried out on a 60 m, 0.32 mm i.d. capillary column internally coated with a 0.25 um film of DB-1. The column was supplied by J&W Scientific (Folsom, CA, USA). A full description of the conditions used for the desorption and analysis of VOC can be found in the literature cited[12]. Reconstructed mass chromatography combined with a detailed knowledge of the elution sequence on DB-1 columns was used for peak identification[12,13]. With this technique more than 300 components with a number of carbon atoms from 4 to 11 present in air samples can be easily identified and quantified in a single GC-MS run[13].

RESULTS AND DISCUSSION

Figure 3 shows the Total Ion Current (TIC) profile obtained by submitting to GC-MS analysis the sample where the highest levels of VOC were measured (Tourmaline Plateau). By combining the information collected in the various sites investigated, a total number of 76 different VOC were unambiguously identified. They are listed in Table 2 together with the concentrations measured in each site. Since enrichment of VOC was carried out at room temperature, reliable values of the mixing ratios were obtained only

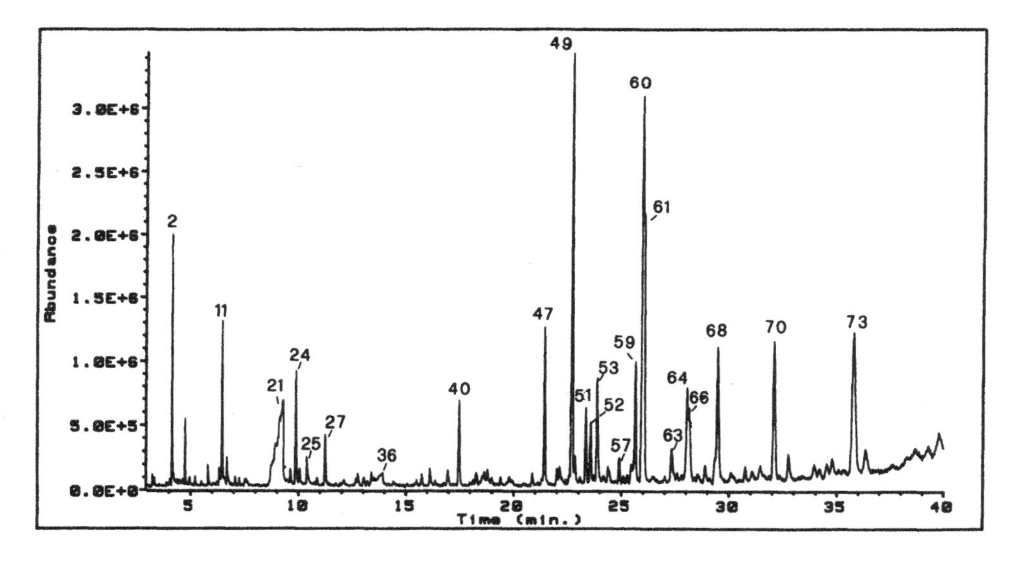

Figure 3 HRGC-MS profile of VOC present at Tourmaline Plateau. For peak identification see Table 2.

Table 2 Concentrations (μg/m³) of VOC measured at 6 different sites located near Terra Nova Bay in Antarctica. (For site identification see Figure 1).

N.	*Compound*	*Ret. time*	*Site 1*	*Site 2*	*Site 3*	*Site 4*	*Site 5*	*Site 6*
1	2-Propenal	4.04	0.35	0.45	0.32	0.60	0.91	0.63
2	Acetone	4.12	16.08	9.62	6.86	8.93	26.82	40.26
3	n-C5	4.55	0.11	–	0.12	0.18	0.57	0.14
4	Isoprene	4.63	0.21	0.21	–	0.36	–	0.48
5	2-Propanol,2-methyl	4.91	0.49	1.52	0.30	0.73	–	–
6	CFC 113	5.21	0.21	0.25	0.13	0.19	0.09	0.13
7	Propanal,2-methyl	5.48	0.27	–	–	0.26	0.38	–
8	2-Propenal,2-methyl	5.79	0.92	0.83	0.61	0.68	1.64	–
9	Furan,2,5-dihydro	6.25	0.25	–	–	–	–	–
10	3-Buten,2-one	6.36	1.28	0.52	0.40	0.84	2.43	2.41
11	Butanal	6.44	9.06	0.72	–	0.90	1.92	1.08
12	2-Butanone	6.63	2.48	1.40	0.85	2.16	4.81	0.98
13	1-Hexene	7.06	0.27	–	–	0.26	0.35	–
14	Furan,2-methyl	7.25	0.29	–	0.20	0.19	0.26	0.47
15	n-C6	7.53	0.14	–	0.13	0.31	0.79	0.21
16	cis-3-Hexene	7.64	0.03	–	–	–	–	–
17	cis-2-Hexene	7.73	0.04	–	–	–	–	–
18	1-Propanol,2-methyl	8.54	–	–	–	1.88	–	–
19	1,1,1-Trichloroethane	8.89	0.51	0.42	0.42	0.51	0.49	0.53
20	2-Butenal	9.20	0.33	0.50	1.12	–	1.16	0.44
21	Acetic acid	(9.5)	27.87	4.02	–	–	8.95	–
22	Benzene	9.62	0.24	–	0.20	0.26	1.27	–
23	CC14	9.87	0.53	0.50	0.54	0.66	0.55	0.59
24	2-Butenal,2-methyl	9.90	4.42	–	–	–	–	–
25	1-Butanol	10.44	1.81	–	–	1.47	1.74	–
26	2-Pentanone	10.82	0.57	0.53	–	0.51	1.20	0.38
27	Pentanal	11.23	2.28	0.69	0.57	0.82	1.89	1.09
28	Ethene, trichloro	11.73	–	–	–	1.04	–	–
29	3-Heptene	12.33	0.03	–	–	–	–	–
30	iso-Heptadiene	12.65	0.21	–	–	–	–	–
31	n-C7	12.73	0.28	–	0.12	0.27	0.56	0.15
32	2-Heptene	13.10	0.07	–	–	–	–	–
33	Formic acid, butyl ester	13.39	0.71	–	–	–	–	–
34	iso-Heptene	13.50	0.07	–	–	–	–	–
35	trans-2-Pentenal	14.05	0.25	–	–	–	–	–
36	Propanoic acid	(14.5)	4.87	–	–	–	–	–
37	Toluene	15.78	0.22	–	–	–	0.58	–
38	cis-2-Pentenal	16.25	0.94	0.52	0.39	0.84	1.29	1.80
39	2-Hexanone	16.98	1.56	0.54	–	1.03	1.15	0.79
40	Hexanal	17.51	3.88	1.61	1.63	1.48	3.81	3.41
41	iso-Octadiene	18.20	0.54	–	–	–	–	–
42	iso-Octene	18.30	0.61	0.32	0.32	0.83	0.76	0.98
43	Butanoic acid	(18.5)	0.84	–	–	0.14	0.60	–
44	iso-Octene	18.80	0.04	–	–	–	–	–
45	n-C8	19.00	0.23	0.52	0.08	0.77	0.69	0.18
46	iso-Octene	19.40	0.24	–	–	–	–	–
47	iso-Heptenal	21.40	5.23	–	–	–	–	–
48	4-Heptanone	22.00	1.31	0.42	–	–	–	–
49	3-Heptanone	22.74	28.18	0.64	–	0.68	–	–
50	2-Heptanone	22.88	1.70	0.50	0.21	0.21	–	–
51	Heptanal	23.33	2.34	0.78	0.74	1.42	2.33	2.32
52	3-Heptanol	23.55	2.32	–	–	–	–	–
53	2-Butoxy,ethanol	23.80	5.06	1.28	1.26	2.18	2.74	14.85
54	2(3H)Furanone,2,3-dihydro,5-methyl	24.35	1.79	–	–	–	–	–
55	n-C9	24.45	0.24	–	0.07	1.02	1.93	0.19

Table 2 *(cont.)* Concentrations (ug/m^3) of VOC measured at 6 different sites located near Terra Nova Bay in Antarctica. (For site identification see Figure 1).

N.	*Compound*	*Ret. time*	*Site 1*	*Site 2*	*Site 3*	*Site 4*	*Site 5*	*Site 6*
56	Pentanoic acid	(24.5)	0.32	–	–	0.05	0.67	–
57	2H-Pyran,2-carboxaldehyde,5,6-dihydro	24.80	1.25	–	–	–	–	–
58	Benzaldehyde	25.53	0.32	0.42	0.23	0.66	1.08	–
59	iso-Octenone	25.70	8.97	–	–	–	–	–
60	iso-Octenone	25.90	41.86	–	–	–	–	–
61	Hexanal,2-ethyl†	26.10	–	–	–	–	–	–
62	Hexanoic acid	(27)	0.37	–	–	0.14	1.01	–
63	5-Hepten,2-one,6-methyl	27.32	3.92	1.05	1.00	3.13	8.11	2.52
64	2-Hexenal,2-ethyl	27.90	3.80	–	–	–	–	–
65	Octanal	28.06	2.70	1.15	0.78	3.90	6.45	2.72
66	2-Octanone,3†-methyl	28.15	–	–	–	–	–	–
67	n-C10	28.91	0.36	–	–	1.32	4.01	0.27
68	1-Hexanol,2-ethyl	29.49	7.21	0.47	0.42	2.22	1.85	1.49
69	Heptanoic acid	(32)	0.14	–	–	0.13	–	–
70	Nonanal	32.10	4.67	3.36	3.42	9.34	8.00	5.25
71	n-C11	32.77	–	–	–	0.40	0.94	–
72	Octanoic acid	(35.5)	0.32	–	–	0.17	–	–
73	Decanal	35.81	6.13	2.98	3.48	9.06	7.29	6.13
74	n-C12	36.36	–	–	–	0.72	–	–
75	Nonanoic acid	(39)	0.58	–	–	0.54	–	–
76	n-C13	39.29	0.26	–	–	–	–	–

for hydrocarbons with a number of carbon atoms larger than ca. 4. The only exception to this general rule is represented by acetic and propionic acid and CCl_4 for which a quantitative collection and full recovery were obtained.

The first observation that can be made by looking at the data shown in Table 2 is that, similarly to what has been observed in the Arctic region, oxygenated components largely dominate the hydrocarbon composition in terms of number of compounds detected and levels reached. Consistently with reduced impact from anthropogenic sources, tiny levels of man-made VOC were found in Antarctica. Among them, only CFC_s, CCl_4 and benzene were present in all sites investigated.

If we focus the attention on components for which reliable values of the mixing ratio were obtained (CCl_4 and benzene), we can see that levels 50% lower than those recorded in the Spitzbergen Islands were detected in Antarctica[4] and they are fully consistent with the ones reported in the literature[14,15]. In addition to benzene, toluene was the only other volatile arene found in Antarctica. It was detected only in the canister sampled at Starr Nunatak. The concurrent presence of large amounts of n-alkanes from C_3 to C_{11} in this sample clearly indicates that the site was directly influenced by anthropogenic emissions. The most likely source of pollution was represented by engine-exhaust emission coming from drilling activities carried out at a remote station for geological studies located some kilometers away from the sampling site.

Although n-alkanes were also detected at Tourmaline Plateau, Edmonton Point, Mount Melbourne and Cape Russel, levels measured were so small that their occurrence in air might be also attributed to biogenic sources. The presence of tiny amounts of isoprene in four of the sites investigated suggets that one of them is probably represented by mosses and lichens as they are the only organisms leaving in Antarctica capable of performing a reduced photosynthetic activity. The possibility of transport can be, in fact, excluded due to short life-time of isoprene in the atmosphere and the long distances of Antarctica from vegetation covered areas.

Although some of the oxygenated compounds listed in Table 2 (such as acetone and 3-butene-2-one) are known to be formed by photochemical degradation of alkanes, alkenes and isoprene, it is unlikely that the low amounts of biomass and reduced photochemical activity occurring in this remote region are capable of producing polar VOC at the levels we measured. This is particularly true for components with number of carbon atoms larger than ca. 8 for which possible precursors and pathways leading to their formation are difficult to envision.

The low photochemical reactivity of the atmosphere is also confirmed by the occurrence of i-octene in all samples. Since this is a highly reactive component never found in more than 300 urban, suburban, rural, forest and remote samples worldwide collected, it can be regarded as a unique constituent of the Antarctic region.

In general, the composition and distribution of oxygenated VOC reported in Table 2 are so not much different from those occurring in the remote areas of the Northern Hemisphere. The organic composition is characterized by high concentrations of acetic acid and aldehydes from propenal to decanal. Only the relatively high abundance of alcohols and alkenals and the lack of esters make possible to distinguish samples from Antarctica from those collected in the Northern Hemisphere. In particular, the lack of esters should be emphasized as it is perfectly on line with the hypothesis that these components are indicators of lower plant (flowers, weeds) emissions not present in Antarctica.

Since vegetation emission and photochemical reactions do not represent a significant source for oxygenated compounds, ocean and/or biogenic activities occurring in it should be responsible for their levels in air. While ocean is known to emit acetic acid[11], it has never been suggested as a possible source for semi-volatile carbonyl components. Based on the results obtained, we believe that marine emission provides a better explanation than long-range transport from vegetation-covered areas for their occurrence in remote islands of the Atlantic[4] and Pacific Ocean[2,3].

As a final comment, it is important to stress here the capabilities afforded by the use of passivated canisters in the collection of polar components in remote areas. In clean environments where low levels of primary and secondary pollution exist, they provide results comparable to those given by adsorption traps. Particularly important is the observation that sticky components, such as free acids, are not significantly depleted by adsorption and/or reactions occurring on their walls.

CONCLUSIONS

Results obtained in Antarctica confirmed the ubiquitous occurrence of oxygenated VOC in the troposphere and the possibility that marine sources might strongly contribute to their atmospheric levels. Since some of these components are sufficiently reactive to participate actively in tropospheric ozone formation, their contribution should be taken into account whenever global changes arising from ozone production are investigated.

Acknowledgments

The financial support of the Italian Antarctic Project is greatly acknowledged.

References

1. P. Ciccioli, E. Brancaleoni, M. Frattoni and A. Cecinato, *Atmos. Environ.*, **27A**, 1891–1901 (1993).
2. Y. Yokouchi, H. Mukai, K. Nakajima and Y. Ambe, *Atmos Environ.*, **24A**, 439–442 (1990).
3. D. Helmig and J. P. Greenberg, *J. Chromatogr.*, **A 677**, 123–132 (1994).
4. P. Ciccioli, A. Cecinato, E. Brancaleoni, A. Brachetti, M. Frattoni and R. Sparapani, *Proceedings of the Sixth European Symposium on the "Physico-Chemical Behaviour of Atmospheric Pollutants"* (G. Angeletti and G. Restelli Edrs., CEC-Air Pollution Research Report EUR 15609/1 EN, Volume 1, Brussels 1994) pp. 549–568.
5. P. Ciccioli, E. Brancaleoni, A. Cecinato, R. Sparapani and M. Frattoni, *J. Chromatogr.*, **643**, 55–69 (1993).
6. P. Ciccioli P., A. Cecinato, E. Brancaleoni, A. Brachetti and M. Frattoni, *Environ. Monit. Assess.*, **31**, 211–217 (1994).
7. R. L. Tanner and B. Zielinska, *Atmos. Environ.*, **28A**, 1113–1120, (1994).
8. J. Arey, A. M. Winer, R. Atkinson, S. M. Ashmann, W. D. Long and C. L. Morrison, *Atmos. Environ.*, **25A**, 1063–1075 (1991).
9. A. M. Winer, J. Arey, R. Atkinson, S. M. Ashmann, W. D. Long, C. L. Morrison and D. Olszyk, *Atmos. Environ.*, **26A**, 2647–2659 (1992).
10. V. A. Isidorov, I. G. Zenkevich and B. V. Ioffe, *Atmos. Environ.* **19**, 1–8 (1985).
11. G. Helas and J. Kesselmeier., *CEC Air Pollution Research Report No. 47* (Angeletti and S. Beilke Edrs., Brussels, 1993, pp. 299–304).
12. P. Ciccioli, A. Cecinato, E. Brancaleoni, M. Frattoni and A. Liberti, *J. High Resolut. Chromatogr. and Chromatogr. Commun.* **15**, 75–84 (1992).
13. P. Ciccioli, E. Brancaleoni, A. Cecinato and M. Frattoni, *Proceedings of the 15the Symposium on Capillary Chromatography* (P. Sandra and G. Devos Eds., Huetig Verlag publisher, Heidelberg, Volume II 1993) pp. 1029–1901.
14. F. Bruner, F. Mangani and M. Maione, *Environ. Monit. Assess.* **31**, 219–224 (1994).
15. H. B. Sing and P. B. Zimmermann, in: *Gaseous Pollutants: Characterization and Cycling* (J. O. Nriagu Edr., John Wiley & Sons, New York, 1992) pp. 177–235.

THE EFFECT OF SEASONAL PACK ICE MELTING ON THE SEA WATER POLYCHLOROBIPHENYL CONTENTS AT GERLACHE INLET AND WOOD BAY (ROSS SEA – ANTARCTICA)*

ROGER FUOCO[a†], STEFANIA GIANNARELLI[a], CARLO ABETE[b], MASSIMO ONOR[a] and MARCO TERMINE[a]

[a]*Department of Chemistry and Industrial Chemistry – University of Pisa Via Risorgimento 35, 56126 Pisa, Italy and* [b]*ICAS – CNR, Via Risorgimento 35, 56126 Pisa, Italy*

(Received 26 February 1999; In final form 12 April 1999)

The effect of seasonal formation/melting process of pack ice on the PCB level of sea water at Gerlache Inlet and Wood Bay (Ross Sea, Antarctica) was investigated during four Italian expeditions, i.e. 1988–89, 1990–91, 1991–92, and 1993–94. Surface sea water samples from Gerlache Inlet and Wood Bay before pack ice melting showed a typical total PCB concentration of 133 pg/l and 120 pg/l, respectively, which increased by a factor of about 1.3 in both sampling sites during pack ice melting. This effect was attributed to the transfer of PCBs contained in the pack ice to sea water, and salinity was used as a tracer to verify this hypothesis. In this respect, pack ice and sea water samples were collected during the 1993–94 Italian expedition, and both salinity (S) and total PCB content were measured. A fairly good agreement was observed between the experimental PCB concentration and the value calculated by the dilution model which was applied to the mixing process between sea and pack ice melting waters. Although this effect seems to be limited in time and space it is nevertheless significant because it happens during summer when biological species have their highest activity.

Keywords: Antarctica (Ross Sea); polychlorobiphenyls; sea water; pack ice

INTRODUCTION

Antarctica plays a very significant role in the global processes of our planet, and observing selected chemical parameters in the Antarctic ecosystem is very important for a better understanding of these processes. The Environmental Pol-

* This paper is a contribution to the Proceedings of the 3rd Italian Antarctica Programme published in Intern. J. Environ. Anal. Chem., 71(3–4), 1998.

† Corresponding author. Fax: +39–050–918260. E-mail: fuoco@dcci.unipi.it

lution Monitoring Project of the Italian Research Programme in Antarctica (PNRA) is aimed at studying these processes by monitoring organic and inorganic pollutants in different environmental components, elucidating the relevant diffusion and distribution processes, and evaluating their change in the short and long terms by sampling those matrices that retain this information, i.e. sediments and ice. Within this framework, our specific interest was concerned with the presence of polychlorobiphenyls (PCBs) in Antarctica. The importance of PCBs is mainly due to the fact that these compounds were used in many industrial applications from around 1930 to 1970, and dumped without any precautions to prevent environmental pollution. Moreover, they are chemically very stable and are able to accumulate in organisms[1,2]. These characteristics are responsible for long residence times in the environment and for toxic effects on biota, thus making the monitoring of PCBs in the environment of prime importance.

This paper shows the most significant findings relevant to four surveys performed at Gerlache Inlet and Wood Bay (Ross Sea – Antarctica) during the 1988–89, 1990–91, 1991–92 and 1993–94 Italian expeditions. In particular, sea water and pack ice samples were collected during the pack ice melting in order to evaluate the role of the seasonal formation/melting process of pack ice on the transfer process of PCBs from the atmosphere to sea water.

EXPERIMENTAL

Reagents

Acetone, n-hexane, dichloromethane Pesticide Grade; Na_2SO_4 RPE-ACS; and Florisil RS (60–100 mesh) were supplied by Carlo Erba (Italy). Na_2SO_4 and Florisil were activated for twelve hours at 450°C, and for four hours at 650°C, respectively, and kept at 130°C for two hours before use. Aroclor 1221, 1232, 1248, 1260 (35 mg/ml) and individual PCB congeners in iso-octane or methanol standard solutions (35 mg/ml) were supplied by AccuStandards (USA). Certified reference material of ten PCBs in iso-octane (CRM 365) were supplied by the European Community-Standards, EC-SM&T Programme (Belgium) (Table I).

Apparatus

A GC mod. 5880A (Hewlett Packard Italiana, Italy), coupled with a mass spectrometric detector (MSD) mod. 5971A (Hewlett Packard Italiana, Italy) and equipped with a cold on-column injection port, was used for the identification of chromatographic peaks and the quantitative determination of PCBs. Chromatographic separation was always performed on a chemically bonded fused silica capillary column MS-5 (Hewlett Packard Italiana, Italy) 0.25 mm I.D., 0.25 μm film thickness, 30 m length, connected to 2 m long deactivated fused silica capillary pre-column 0.32 mm I.D. The chromatographic conditions were Ti = 50°C or 80°C for n-hexane or iso-octane solution, respectively, and isothermal for 2 min, then 15°C/min up to 180°C and isothermal for 6 min, 4°C/min up to 220°C and isothermal for 2 min, 5°C/min up to 280°C and isothermal for 25 min; the carrier gas was helium at 190 KPa.

TABLE I PCBs in iso-octane CRM 365

IUPAN No.	*Content, μg/g*
PCB8	11.4 ± 0.3
PCB 20	15.2 ± 0.9
PCB 28	24.8 ± 1.1
PCB 35	14.3 ± 0.8
PCB 52	14.8 ± 0.6
PCB 101	14.4 ± 0.6
PCB 118	14.9 ± 0.8
PCB 138	8.6 ± 0.6
PCB 153	14.2 ± 0.6
PCB 180	15.2 ± 0.6

Sampling area and sampling techniques

Sea water samples were collected at about 0.5 m from the surface by a teflon pumping system at Gerlache Inlet (sampling stations Nos. 1, 2, 3 and 4) and Wood Bay (sampling station No. 5) (Figure 1). Before pack-ice melting a hole in the ice was made by a manual drilling system. Pack ice was collected during this operation, and sea water was collected underneath the inner surface of the pack-ice, at a depth of about 0.5 m. The samples were either immediately extracted with n-hexane in the clean-room facility at the Italian Base or stored at –20°C and sent to Italy and analysed in our laboratory. All samples were stored in pre-cleaned 20 1 stainless steel containers.

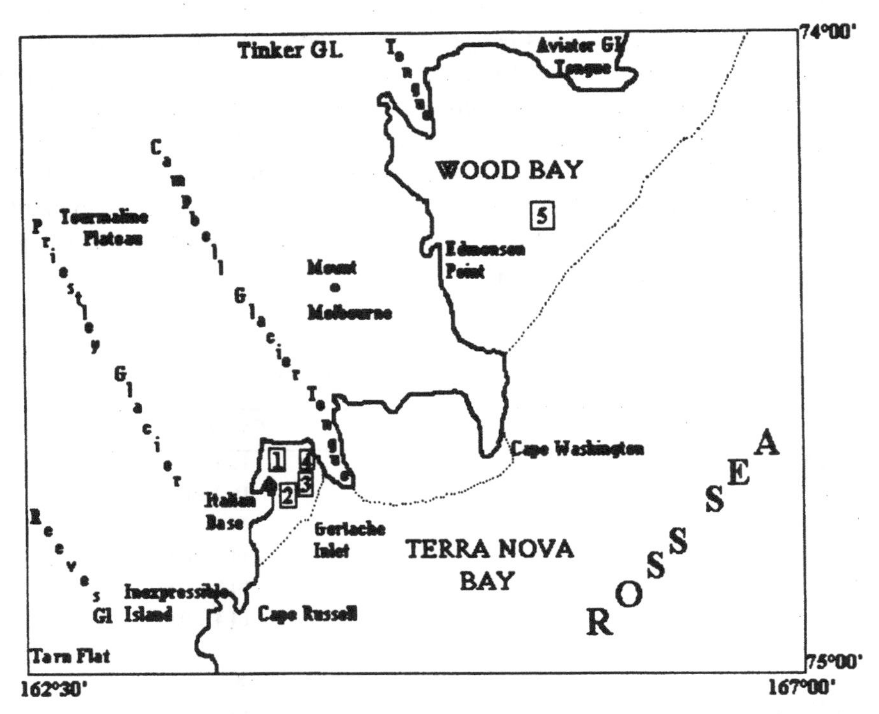

FIGURE 1 Location of the sea water sampling stations before and during pack ice melting at Gerlache Inlet, (1988–89, 1990–91, and 1991–92 Italian expeditions) and Wood Bay (1993–94 Italian expedition)

Analytical procedures

The analytical procedure for total PCB determination in sea water has been described elsewhere[3,4]. The same procedure was also applied to pack ice samples. In short, the samples were liquid-liquid extracted either by mechanical shaking in a separator funnel or by a custom-made high efficiency extraction system[5] which allows the use of about 10 ml of n-hexane to extract 10 litres of water. In both cases the extract was cleaned up on a Florisil column (50×10 mm), from which PCBs were selectively eluted with a suitable volume of n-hexane, typically 10 ml. Right before the analysis, the eluate was concentrated at a suitable volume, and analysed by gas chromatography-mass spectrometry.

Total PCB concentration was always obtained by measuring the individual concentration of 60–70 congeners. In particular, PCB congeners were firstly identified by GC/MSD on a standard solution of several Aroclors (i.e., 1221,

1232, 1248 and 1260). The relative retention time (RRT) for each identified congener was then calculated by using one or more internal standards (ISs). RRTs were finally applied for chromatographic peak assignment of real samples analysed by either GC/MSD or GC/ECD. Experimental response factors (RFs) were generally obtained for a limited number of selected PCB congeners (i.e., IUPAC Nos. 8, 20, 28, 35, 52, 101, 118, 138, 153, 180) and for the IS, in a selected concentration range. Relative response factors (RRFs) to the IS were then calculated and used in turn to calculate the RRFs for all congeners by extrapolating the values reported by Mullin[6]. If an MS detector in the Selected Ion Monitoring mode is used, at least three ions should be selected: one as target and two as qualifiers.

Analytical quality control and quality assurance

Analytical quality control and quality assurance programs were run in the laboratory during sample analysis in order to get reliable data. These programs were based on the use of standard solutions suitably prepared and stored, and spiked samples, since reference sea water samples with a certified PCB content are unfortunately not available. Table I shows the concentration of ten PCB congeners which are present in the certified reference material CRM 365, while Figure 2 shows a typical control chart for the determination of PCB153 in CRM 365, as obtained during the analysis of sea water and pack ice samples.

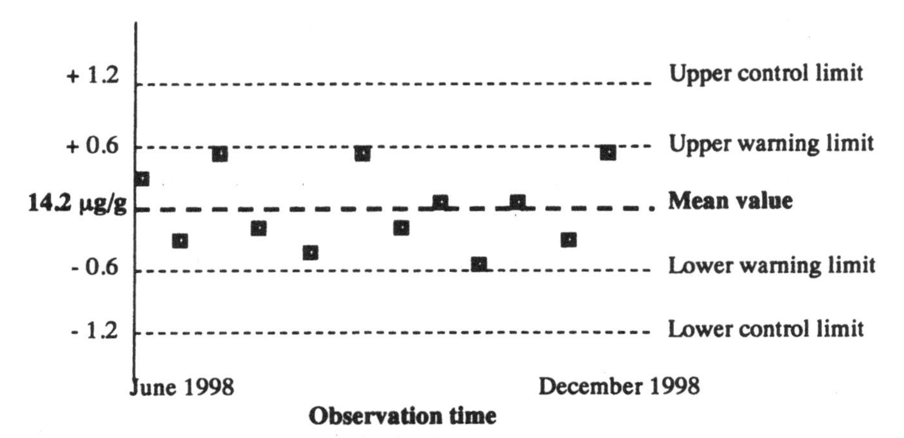

FIGURE 2 Working analytical quality control chart for PCB153 determination in iso-octane certified standard solution CRM 365

RESULTS AND DISCUSSION

Table II shows the total PCB content of surface sea water samples collected during four Italian expeditions in Antarctica before and during pack ice melting. The concentration ranges refer to samples collected in all the sampling stations for each expedition. In particular, before pack ice melting a mean value of 133 and 120 pg/l was observed at Gerlache Inlet and Wood Bay, respectively. These concentrations increased during pack ice melting by a factor of about 1.3 in both sampling sites (180 pg/l at Gerlache Inlet, 160 pg/l at Wood Bay). In order to evaluate whether this effect might be attributed to the transfer of PCBs contained in the pack ice to sea water, salinity was used as a tracer. In fact, by measuring salinity (S) and total PCB content in both pack ice and sea water samples it should be possible to evaluate the mixing process between sea and pack ice melting waters which affects the superficial layer of sea water for a limited period of time.

TABLE II Effect of pack ice melting on the total PCB content in surface sea water samples collected at Gerlache Inlet and Wood Bay, Ross Sea-Antarctica (The mean value is reported in brackets.)

Expedition	*Location*	*Sampling stations*	*Tolal PCBs, pg/l*	
			Before pack ice melting	*During pack ice melting*
1988–89	Gerlache Inlet	4	100 – 160 (120)	150–180 (160)
1990–91	Gerlache Inlet	1	140	180
1991–92	Gerlache Inlet	4	90 –180 (140)	160 – 2<30 (205)
1993–94	Wood Bay	2	120	160

The dilution factor of each sample can be described by the following equation:

$$\alpha = \frac{V_{sw}}{V_{sw} + V_{pk}} = \frac{(S_i - S^0_{pk})}{(S^0_{sw} - S^0_{pk})} \tag{1}$$

where:

V_{sw} = volume of sea water; V_{pk} = volume of water from pack ice melting; S_i = sample salinity; S^0_{pk}= pack ice salinity; S^0_{sw} = sea water salinity before pack ice melting.

Before pack ice melting the variables have the following values:

$$V_{pk} = 0 \qquad S_i = S^0_{sw} \qquad \alpha = 1$$

and during pack ice melting:

$$V_{pk} > 0 \qquad S_i < S_{sw}^0 \qquad \alpha < 1$$

If this model holds, and in the absence of other sources of PCBs in the studied area, the total PCB concentration of each sample can be calculated by the following equation:

$$C_i = C_{pk}^0 - (C_{pk}^0 - C_{sw}^0)\alpha_i \tag{2}$$

where:

C_i = total PCB concentration of the sample; C^0_{pk} = total PCB concentration of pack ice sample; C^0_{sw} = total PCB concentration of sea water sample before pack ice melting.

Figure 3 shows four depth profiles of salinity obtained at Wood Bay from 27 November 1993 to 28 January 1994. These results show that:

- on 27 Nov 1993 the pack ice melting process had not yet begun;
- on 15 Dec 1993 the decrease in the salinity in the surface layer of sea water due to pack ice melting was clearly evident;
- on 9 Jan 1994 the largest variation of salinity was observed and was confined in a layer of sea water about 80 m thick.

Since the scientific activities at the Italian Base had to be concluded on 31 Jan 1994, it was not possible to observe the evolution of the salinity profile for longer. Nevertheless, it was known that the formation of new ice patches on the sea surface in the studied area began in the middle of February, thus pack ice melting was over at that time. Consequently, the time lag during which pack ice melting affected the PCB content of sea water was about two months, from 15 December to 15 February. In order to verify this hypothesis, the total PCB content was determined in sea water and pack ice samples, before and during pack ice melting, and compared with the calculated value according to equation (2). Table III shows the results for the samples collected at Wood Bay during the 1993–94 expedition. In particular, the pack ice showed a salinity of 6 ‰, whereas the salinity of sea water samples collected at the surface before (profile of 27 Nov 1993, Figure 3) and during (profile of 9 Jan 1994, Figure 3) pack ice melting was 34.7 ‰ and 33.6 ‰, respectively. The corresponding value of the dilution factor (α) calculated by equation (1) was 0.96. On the other hand, the total PCB content in both sea water and pack ice samples collected on 27 November 1993 was 120 and 750 pg/l, respectively. This shows a quite high total PCB content in pack ice samples. Finally, the total PCB content in sea water sample collected on 9 January 1994 was 160 pg/l. A fairly good agreement between the total PCB content calculated by equation (2) (145 pg/l), and the experimental one (160 pg/l) was observed. According to the value of the standard deviation, as obtained in five replicate measurements, the increase in PCB con-

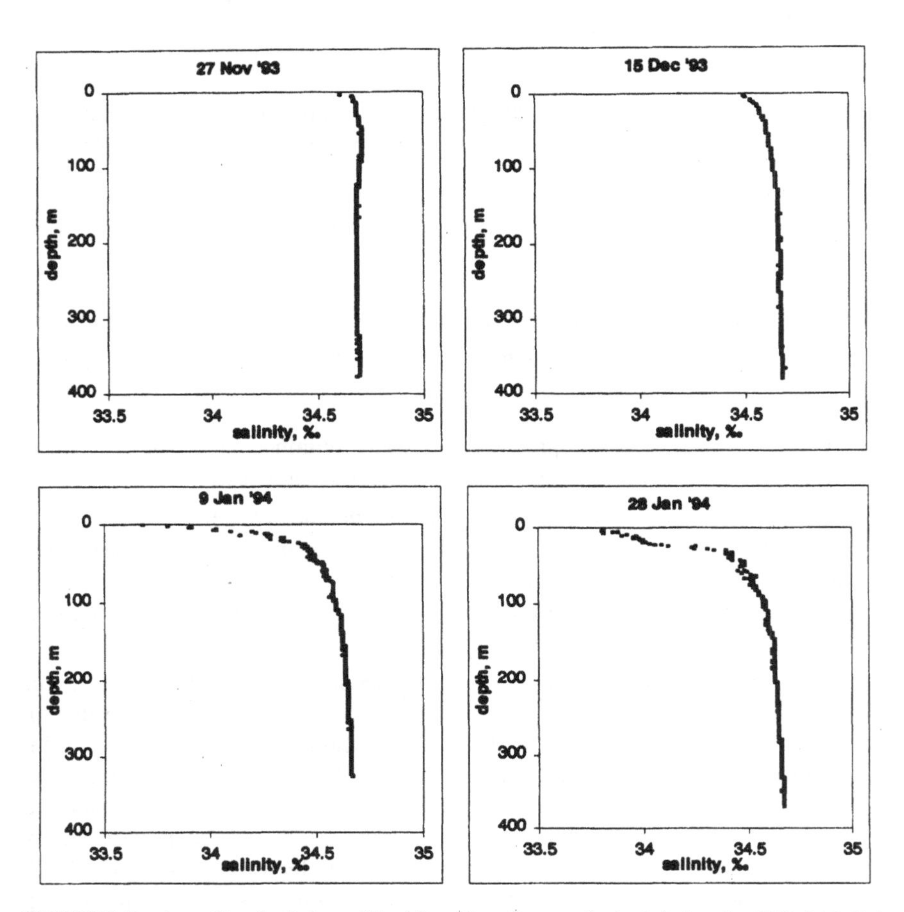

FIGURE 3 Depth profile of salinity at Wood Bay, Ross Sea, as obtained during the 1993–94 Italian expedition

centration was statistically significant at a 90% confidence level, and should be attributed to pack ice melting if we assume that there was no local source of PCB contamination.

TABLE III Total PCB content and salinity of sea water and pack ice samples collected at Wood Bay during the 1993–94 Italian expedition

Sample	*Salinity, g % o*	α	*Total PCBs, pg/l*
Sea water, before pack ice melting	34.7	1.00	120
Pack ice	6.0		750
Sea water, during pack ice melting	33.6	0.96	Experimental 160 Calculated (*) 145

(*) The calculated value was obtained by equation (2), see text.

This increase might be explained considering that ice acts, during its formation, as an accumulator which traps atmospheric particulate material that is realeased, along with the pollutants adsorbed on it, to the sea water during melting. Although this effect seems to be limited in time and space, and the PCB content only increases by about 30%, it is nevertheless significant because it happens during summer time when biological species have their highest activity.

Acknowledgements

This work was financially supported by the Italian Research Programme in Antarctica (PNRA).

References

[1] M.D. Erickson, *Analytical Chemistry of PCBs,* (Butterworth Publishers Stonehouse, MA, USA, 1986), p. 24.
[2] J. Albaiges (editor), *Environmental Analytical Chemistry of PCBs,* (Gordon and Breach, Reading, UK, 1993), p. 40.
[3] R. Fuoco, M. P. Colombini and C. Abete, *Intern. J. Environ. Anal. Chem.*, **55**, 15–25 (1994).
[4] R. Fuoco, M. P. Colombini, C. Abete and S. Carignani, *Intern. J. Environ. Anal. Chem.*, **61**, 309–3181 (1995).
[5] L. Zoccolillo and R. Fuoco, *private communication.*
[6] M.D. Mullin, C.M. Pochini, S. McCrindle, M. Romkes, S.H. Safe, L.M. Safe, *Environ. Sci. Technol.*, **18**, 468–476 (1984).

VARIATIONS OF MERCURY AND SELENIUM CONCENTRATIONS IN *ADAMUSSIUM COLBECKI* AND *PAGOTHENIA BERNACCHII* FROM TERRA NOVA BAY (ANTARCTICA) DURING A FIVE YEARS PERIOD

V. MINGANTI[1], R. CAPELLI[1], F. FIORENTINO[2], R. DE PELLEGRINI[1] and M. VACCHI[3]

[1]*Università di Genova, Istituto di Analisi e Tecnologie Farmaceutiche ed Alimentari, Via Brigata Salerno (ponte), 16147 Genova, Italy;* [2]*AQUATICA, Via F.lli Canale 22/26, 16132 Genova, Italy;* [3]*ICRAM (Istituto Centrale per la Ricerca Scientifica e Tecnologica Applicata al Mare), Via L. Respighi 5, 00197 Roma, Italy*

Pagothenia bernacchi and *Adamussium colbecki* specimens have been collected in Terra Nova Bay, Ross Sea, Antarctica, from 1987 to 1992. Total and organic mercury and selenium concentrations were measured in the muscular tissue of *P. bernacchii* and in the soft parts of *A. colbecki*. Sampling, samples preparation, and analysis were the same for all the specimens, and the temporal variation in concentrations are discussed. Large variations for total and organic mercury contents are found in *A. colbecki*, while *P. bernacchii* shows significant interannual differences in selenium concentration. However, none of these differences reflect a consistent temporal trend. Medians obtained during the five years in 109 specimens of *P. bernacchii* are: 0.60 µg/g dry weight for total mercury, 78% for the percentage of organic mercury, and 3.0 µg/g dry weight for selenium. Values found in 117 specimens of *A. colbecki* are: 0.17 µg/g dry weight for total mercury, 28% for the percentage of organic mercury, and 9.1 µg/g dry weight for selenium.

KEY WORDS: Antarctica, marine organisms, mercury, selenium.

INTRODUCTION

The scallop *Adamussium colbecki* and the bony fish *Pagothenia bernacchii* are two widely distributed Circum-Antarctic species, and they both can be easily sampled in the neighbourhood of the Italian Base in Terra Nova Bay, Ross Sea.

A. colbecki is a filter feeding bivalve able to move, in some degree, on the substratum[1]. It is usually found on soft and mixed bottoms, typically between 8 and 75 meters water depth in the area in study[2]. *P. bernacchii* is a predator, feeding on small benthic invertebrates such as *A. colbecki*[3]. It represents the most abundant fish in the area, and it is typically caught between 0 and 100 meters[4]. The two species belong to quite different trophic levels and they both have the features of the typical bioindicator[5]. More information on biology and detailed descriptions of these species are reported by Fisher and Hureau[6].

During five Expeditions carried out in the framework of the Italian National Program of Researches in Antarctica (PNRA), between 1987 and 1992, 117 specimens of *A. colbecki* and 109 of *P. bernacchii* were collected and analysed for organic mercury, total mercury, and selenium concentrations. The amount of data acquired, which spans over five years, permitted to evaluate temporal variations of these elements in the Antarctica.

EXPERIMENTAL

Samples of *P. bernacchii* and *A. colbecki* were collected in the Terra Nova Bay, Ross Sea (Antarctica) during the Antarctic Expeditions organised by Italy during the five austral summers: 1987–88 to 1991–92.

After sampling, the specimens were stored at –25°C until analysis in the laboratory. The *A. colbecki* specimens were measured (length of the shell along the dorso-ventral axis from the umbone region), weighed as whole and soft parts and sexed. The *P. bernacchii* individuals were measured, weighed and sexed.

The soft parts of scallops and the muscular tissue of fishes were freeze-dried and homogenised. An aliquot (0.2–0.5 g dry weight) of homogenate was mineralized by 3–8 ml of 65% nitric acid (Merck Suprapur) in Pyrex volumetric flasks, equipped with air cooled condersers.

Total mercury was determined by cold vapour atomic absorption spectrometry (CVAAS, Perkin-Elmer Mod. 560) by reduction with 10% (w/v) tin(II) chloride dihydrate in 10% (v/v) sulphuric acid, using the preconcentration over gold technique, using an accessory built in our laboratory.

Selenium was determined with the hydride generation (HGAAS) method by reduction with sodium borohydride. A Perkin Elmer 1100B AAS equipped with the MHS-20 accessory was used.

Organic mercury was determined on a second aliquot of the homogenate, after extraction in toluene and back-extraction in a cysteine solution. The sample (0.2–0.5 g dry weight) was mixed with 2 ml 47% hydrobromic acid and 12 ml of toluene, shaken manually for 5 minutes and centrifuged. The toluene (10 ml) was transferred into another polyethylene tube and 5 ml of 1% (w/v) L-cysteine solution were added. After shaking for 5 minutes (an antifoam agent can be used if any foam appears), the samples are centrifuged and the cysteine solution is recovered. Aliquots of the aqueous phase were analysed with the same equipment utilized for the total mercury determination. The reducing solution was 1 ml of 50% (w/v) tin (II) chloride-di-hydrate and 10% (w/v) cadmium chloride monohydrate solution and 4 ml of 10% (w/v) sodium hydroxide solution. Reduction is carried out at 70°C.

Calibration is carried out by the standard additions method. "Blank" samples are processed together with real samples.

Accuracy of analytical determinations was verified using Standard Reference Materails obtained from the IAEA/Monaco Laboratory or from the National Research Council, Canada. A summary of results obtained during a period of about four years (42 months) is reported in Figure 1. Fluctuations are evident but there are no trends which can explain the differences found.

RESULTS AND DISCUSSION

Results concerning mercury (total and organic) and selenium concentrations in the muscular tissue of *P. bernacchii* collected during 1987–88[7], 1988–89[8], and 1989–90[9],

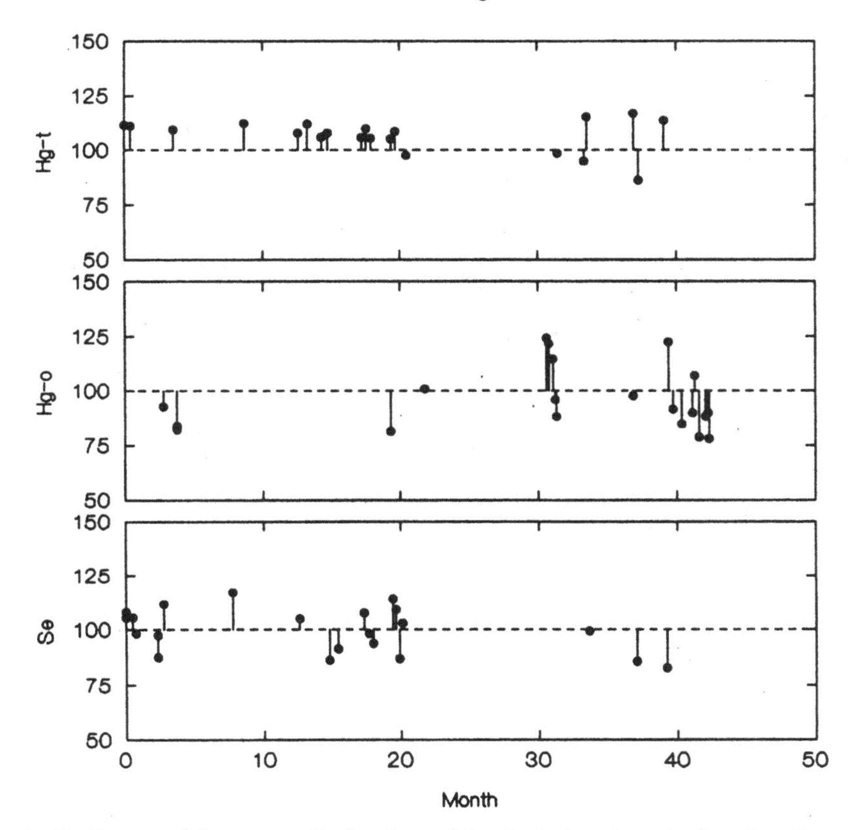

Figure 1 Quality control for mercury (total and organic) and selenium determinations from January 1990 to July 1993. Standard Reference Materials used are: IAEA/Monaco Laboratory MA-M-2 (Mussel tissue), and National Research Council, Canada, TORT-1 (Lobster hepatopancreas). Results are shown as percentage of certified or reference values.

and in the soft parts of *A. colbecki* collected during 1987–88[7], and 1988–89[10] have been already published. Latest unpublished data are reported in Tables 1–5. Distributions of data obtained, for all the five years, are summarised as "Box and Whisker" plots[11] in Figures 2–4. Differences in concentrations between the years are tested using one-way analysis of variance[12].

The differences for total mercury concentration among the years examined for *A. colbecki*, shown in Figure 2, are statistically significant ($F = 11.06$, $p < 0.001$). In 1987–88 and 1989–90, concentrations are lower (mean 0.14 ± 0.04 μg/g dry weight, n = 41) than in all other years (mean 0.20 ± 0.06 μg/g dry weight, n = 76).

It has to be noted that mercury concentration depends on the size of the specimen, as shown in Figure 5. The correlation coefficients (r) between total mercury concentration and size are 0.536 ($n = 111$, $p < 0.001$) and 0.553 ($n = 109$, $p < 0.001$) for *A. colbecki* and *P. bernacchii*, respectively. However, the differences in mercury concentration described for *A. colbecki* are not due to the size of specimens analysed. In fact, the same findings are observed by comparing specimens homogeneous in size. *P. bernacchii* shows no significant differences among the years, and the mean value, for all samples, is 0.64 ± 0.31 μg/g dry weight (n = 109).

Table 1 Results obtained for muscular tissue of *Pagothenia bernacchii* collected during the 1990–91 Expedition.

Samp.	*Length cm*	*Weight g*	*Sex*	*FW/DW*	*Hg-t µg/g DW*	*Hg-o µg/g DW*	*Hg-o%*	*Se µg/g DW*
PABE22	22.0	160.0	F	5.16	0.476	0.408	85.7	3.30
PABE23	22.5	160.0	F	5.25	0.535	0.496	92.7	3.24
PABE24	23.0	163.0	U	5.31	0.515	0.517	100.0	2.98
PABE25	24.5	211.0	F	5.26	0.600	0.495	82.5	3.52
PABE26	25.5	228.0	F	4.85	0.394	0.330	83.8	2.81
PABE27	19.5	105.0	F	6.36	0.541	0.541	100.0	3.96
PABE28	22.0	149.0	F	5.27	0.310	0.214	69.0	3.28
PABE29	28.0	280.0	F	6.02	1.210	0.961	79.4	3.01
PABE30	22.5	166.0	F	5.30	0.480	0.421	87.7	2.86
PABE31	24.5	216.0	F	5.09	0.660	0.510	77.3	4.16
PABE32	23.0	189.0	F	5.28	0.485	0.386	79.6	3.01
PABE33	27.0	235.0	F	6.12	1.080	0.820	75.9	4.11
PABE34	20.5	123.0	F	4.84	0.310	0.170	54.8	2.76
PABE35	24.5	239.0	F	5.17	0.990	0.680	68.7	3.00
PABE36	23.5	185.0	M	5.47	0.975	0.901	92.4	3.37
PABE37	25.5	249.0	F	5.54	0.600	0.236	39.3	3.44
PABE38	25.5	240.0	F	5.39	0.920	0.790	85.9	3.96
PABE39	26.5	263.0	F	5.61	0.870	0.790	90.8	2.59
PABE40	24.5	226.0	F	5.44	0.740	0.594	80.3	4.38
PABE41	25.5	241.0	F	5.37	0.420	0.370	88.1	2.72
PABE42	21.0	146.0	M	5.41	0.420	0.392	93.3	2.40
PABE43	22.0	139.0	M	5.11	0.500	0.433	86.6	3.28
PABE44	30.5	430.0	F	5.87	1.620	1.434	88.5	2.25
PABE45	22.5	178.0	F	5.19	0.660	0.518	78.5	3.92
PABE46	25.0	202.0	M	5.75	1.100	0.942	85.6	3.10
PABE47	22.0	166.0	F	5.06	0.520	0.350	67.3	3.20
PABE48	21.5	136.0	M	5.23	0.590	0.419	71.0	2.70
PABE49	20.0	101.0	F	5.19	0.370	0.264	71.4	2.52
PABE50	21.0	136.0	U	5.48	0.340	0.205	60.3	2.24

The following is reported for each sample: sample code (Samp.), total length in cm; weight in g; sex (F = Female, M = Male, U = Undetermined); fresh weight/dry weight ratio (FW/DW); concentration of total mercury (Hg-t), organic mercury (Hg-o), and selenium (Se) in µg/g DW; percentage of organic mercury (Hg-o%).

Table 2 Results obtained for muscular tissue of *Pagothenia bernacchii* collected during the 1991–92 Expedition.

Samp.	*Length cm*	*Weight g*	*Sex*	*FW/DW*	*Hg-t µg/g DW*	*Hg-o µg/g DW*	*Hg-o%*	*Se µg/g DW*
PABE51	21.0	116.0	F	5.21	0.879	0.452	51.4	2.59
PABE52	21.2	163.7	F	5.11	0.322	0.307	95.3	2.90
PBAE53	22.1	137.6	M	4.31	0.342	0.341	99.7	3.40
PABE54	20.0	124.0	M	5.06	0.186	0.162	87.1	3.24
PABE55	20.4	119.3	M	5.09	0.199	0.154	77.4	2.59
PABE56	22.0	146.3	F	4.98	0.461	0.425	92.9	2.50
PABE57	16.2	45.6	M	4.76	0.242	0.242	100.0	2.75
PABE58	22.6	168.3	M	5.58	0.879	0.826	94.0	1.82
PABE59	23.2	291.0	F	5.40	0.806	0.643	79.8	2.63
PABE60	22.5	237.5	F	5.13	0.389	0.335	86.1	3.23
PABE61	28.6	403.8	F	4.60	1.344	1.344	100.0	2.07
PABE62	27.5	332.2	U	7.48	0.838	0.824	98.3	2.07

The following is reported for each sample: sample code (Samp.), total length in cm; weight in g; sex (F = Female, M = Male, U = Undetermined); fresh weight/dry weight ratio (FW/DW); concentration of total mercury (Hg-t), organic mercury (Hg-o), and selenium (Se) in µg/g DW; percentage of organic mercury (Hg-o%).

Table 3 Results obtained for soft parts of *Adamussium colbecki* collected during the 1989–90 Expedition.

Samp.	*Length mm*	*Weight g*	*Sex*	*FW/DW*	*Hg-t µg/g DW*	*Hg-o µg/g DW*	*Hg-o%*	*Se µg/g DW*
AC1/7	67.0	12.1	M	7.18	0.250	0.079	31.6	7.46
AC1/8	82.0	20.3	F	6.33	0.180	0.078	43.3	5.28
AC1/9	56.0	9.2	M	6.40	0.148	0.064	43.2	7.00
AC1/10	69.0	12.4	F	5.99	0.172	0.064	37.2	10.88
AC1/11	73.0	14.5	U	5.62	0.132	0.100	75.8	11.21
AC1/12	76.0	14.7	F	6.87	0.158	0.121	76.6	9.74
AC1/13	59.0	7.5	F	6.06	0.087	0.043	75.4	10.70
AC1/14	68.0	9.9	U	6.81	0.148	0.086	58.1	9.86
AC1/16	82.0	17.0	F	6.57	0.178	0.098	55.1	12.96
AC1/17	58.0	7.9	F	5.11	0.073	0.029	39.7	8.67
AC1/18	67.0	17.8	U	5.19	0.060	0.046	76.7	9.17
AC1/20	64.0	7.3	F	5.30	0.095	0.046	48.4	13.15

The following is reported for each sample: sample code (Samp.), length of the shell in mm; weight of the soft parts in g; sex (F = Female, M = Male, U = Undetermined); fresh weight/dry weight ratio (FW/DW); concentration of total mercury (Hg-t), organic mercury (Hg-o), and selenium (Se) in µg/g DW; percentage of organic mercury (Hg-o%).

Table 4 Results obtained for soft parts of *Adamussium colbecki* collected during the 1990–91 Expedition.

Samp.	*Length mm*	*Weight g*	*Sex*	*FW/DW*	*Hg-t µg/g DW*	*Hg-o µg/g DW*	*Hg-o%*	*Se µg/g DW*
AC20	71.6	25.6	M	5.14	0.168	0.047	28.0	13.86
AC21	76.0	26.4	F	4.87	0.205	0.064	31.2	7.96
AC22	77.7	23.5	M	5.53	0.310	0.078	25.2	11.96
AC23	72.4	21.9	F	5.32	0.187	0.050	26.7	10.86
AC24	81.4	26.9	F	5.33	0.230	0.070	30.4	13.30
AC25	76.5	23.6	M	4.07	0.372	0.120	32.3	14.90
AC26	67.3	19.3	F	4.80	0.106	0.030	28.3	9.53
AC27	72.0	23.0	F	4.91	0.185	0.057	30.8	9.10
AC28	66.0	19.9	F	4.54	0.166	0.061	36.7	10.80
AC29	71.0	22.9	M	4.79	0.160	0.044	27.5	9.05
AC30	71.1	23.8	M	5.05	0.176	0.057	32.4	11.12
AC31	71.6	22.0	F	5.03	0.154	0.055	35.7	11.32
AC32	70.1	24.0	M	5.19	0.205	0.068	33.2	9.60
AC33	72.2	25.2	M	5.54	0.198	0.067	33.8	11.96
AC34	72.3	22.6	F	5.19	0.166	0.047	28.3	8.48
AC35	70.6	21.6	F	5.13	0.214	0.077	36.0	14.61
AC36	78.5	27.9	F	4.64	0.177	0.045	25.4	10.65
AC37	72.0	19.5	F	5.31	0.276	0.105	38.0	10.39
AC38	68.0	20.6	F	5.13	0.202	0.056	27.7	8.81
AC39	—	25.2	F	5.01	0.164	0.052	31.7	8.38
AC40	67.3	19.6	F	4.81	0.127	0.038	29.9	6.78
AC41	75.3	24.1	F	5.05	0.188	0.070	37.2	8.44
AC42	83.0	33.1	F	5.35	0.157	0.066	42.0	8.44
AC43	79.0	24.3	F	5.15	0.242	0.078	32.2	7.95
AC44	74.2	25.5	F	4.73	0.140	0.054	38.6	9.28

The following is reported for each sample: sample code (Samp.), length of the shell in mm; weight of the soft parts in g; sex (F = Female, M = Male); fresh weight/dry weight ratio (FW/DW); concentration of total mercury (Hg-t), organic mercury (Hg-o), and selenium (Se) in µg/g DW; percentage of organic mercury (Hg-o%).

Percentages of organic mercury are reported in Figure 3. *A. colbecki* shows significant differences among the years considered, with a maximum of 55 ± 17% (n = 12) in

Table 5 Results obtained for soft parts of *Adamussium colbecki* collected during the 1991–92 Expedition.

Samp.	*Length mm*	*Weight g*	*Sex*	*FW/DW*	*Hg-t µg/g DW*	*Hg-o µg/g DW*	*Hg-o%*	*Se µg/g DW*
AC75	74.0	17.5	F	6.84	0.280	0.089	31.8	11.31
AC76	71.0	18.0	F	7.04	0.272	0.074	27.2	7.54
AC77	72.0	17.3	F	7.24	0.180	0.048	26.7	8.07
AC78	—	18.3	M	6.43	0.126	0.038	30.2	6.35
AC79	76.0	21.2	F	7.03	0.362	0.084	23.2	6.41
AC80	84.0	21.5	F	7.14	0.238	0.069	29.0	6.85
AC81	84.0	23.0	F	7.35	0.290	0.070	24.1	6.94
AC82	70.0	20.4	M	6.53	0.202	0.074	36.6	7.91
AC83	75.0	16.4	F	6.97	0.307	0.090	29.3	8.59
AC84	79.0	24.8	F	6.70	0.186	0.032	17.2	4.91
AC85	76.0	19.8	F	7.08	0.217	0.067	30.9	3.67
AC86	77.0	15.8	F	7.31	0.292	0.101	34.6	7.78
AC87	75.0	18.9	F	6.95	0.212	0.038	17.9	6.83
AC88	79.0	18.9	F	7.20	0.234	0.056	23.9	7.25
AC89	68.0	11.8	F	6.24	0.175	0.042	24.0	6.89
AC90	74.0	19.6	M	6.34	0.132	0.035	25.6	5.85
AC91	—	17.9	M	6.70	0.169	0.035	20.7	6.83
AC92	72.0	18.2	F	6.69	0.236	0.082	34.8	9.43
AC93	66.0	18.4	M	7.00	0.248	0.055	22.2	7.74
AC94	81.0	20.0	F	6.86	0.210	0.067	31.9	4.83
AC95	81.0	30.1	M	6.70	0.146	0.034	23.3	6.78
AC96	75.0	17.1	F	6.27	0.210	0.058	27.6	7.69
AC97	72.0	19.1	U	6.75	0.174	0.058	33.3	8.92
AC98	75.0	17.9	F	6.94	0.183	0.066	36.1	6.94
AC99	—	18.4	M	6.08	0.181	0.082	45.3	8.52
AC100	—	15.5	F	5.89	0.145	0.038	26.2	6.10

The following is reported for each sample: sample code (Samp.), length of the shell in mm; weight of the soft parts in g; sex (F = Female, M = Male, U = Undetermined); fresh weight/dry weight ratio (FW/DW); concentration of total mercury (Hg-t), organic mercury (Hg-o), and selenium (Se) in µg/g DW; percentage of organic mercury (Hg-o%).

1989–90, and a minimum of 19 ± 16% (n = 29) in 1987–88. For 1988–89, 1990–91, and 1991–92 no significant differences are observed, and the mean value is 30 ± 6% (n = 76). *P. bernacchii*, shows in 1991–92 a percentage of organic mercury (mean 88 ± 14%, n = 12) higher than those in other years (mean 75 ± 16%, n = 97).

Mercury, as total concentration and as organic fraction, is higher in *P. bernacchii* than in *A. colbeckii*. This is in agreement with the already known magnification of mercury in the aquatic environment[13–15], being the *P. bernacchii* at a higher trophic level than *A. colbecki*. Considering all the differences among the sampling years observed, it can be noted that the two species respond differently. It is not surprising that larger variations occur in *A. colbecki*, since this species is more sensitive to environmental changes[13,16].

Selenium too, reported in Figure 4, exhibits significant differences, both for *A. colbecki* ($F = 11.74$, $p < 0.001$) and for *P. bernacchii* ($F = 18.60$, $p < 0.001$). For *A. colbecki*, selenium concentrations in 1991–92 (mean 7.19 ± 1.54 µg/g dry weight, n = 26) are lower than in the other years (mean 10.31 ± 2.27 µg/g dry weight, n = 91). For *P. bernacchii* the higher concentration is detected in 1987–88 (mean 4.13 ± 0.95 µg/g dry weight, n = 25), followed by 1990–91 (mean 3.17 ± 0.58 µg/g dry weight, n = 29). In 1988–89, 1989–90, and 1991–92 no differences are evident and the mean content is 2.57 ± 0.67 µg/g dry weight (n = 46).

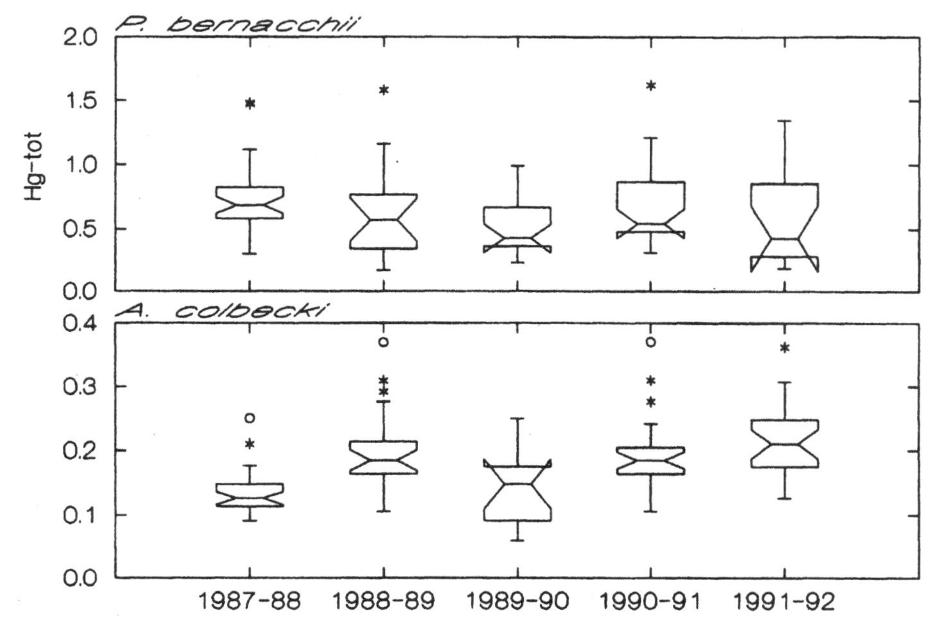

Figure 2 Comparison of mercury concentrations, expressed as μg/g dry weight, in muscular tissue of *P. bernacchii* (top), and in soft parts of *A. colbecki* (bottom), observed from 1987 to 1992. Outliers have been plotted as * (> 1.5 times the interquartile range), o (> 3 times the interquartile range).

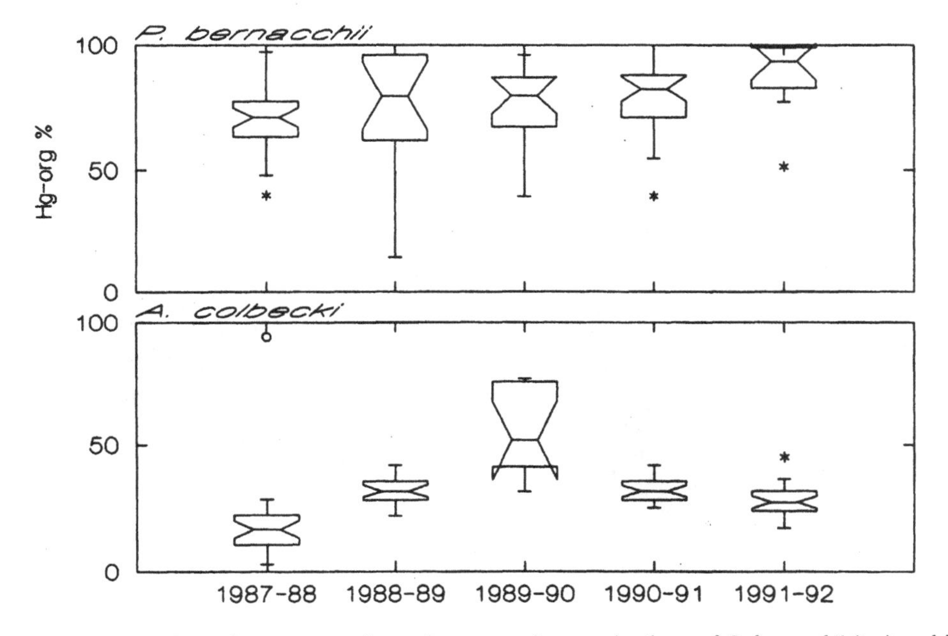

Figure 3 Comparison of percentages of organic mercury, in muscular tissue of *P. bernacchii* (top), and in soft parts of *A. colbecki* (bottom), observed from 1987 to 1992. Outliers have been plotted as * (> 1.5 times the interquartile range), o (> 3 times the interquartile range).

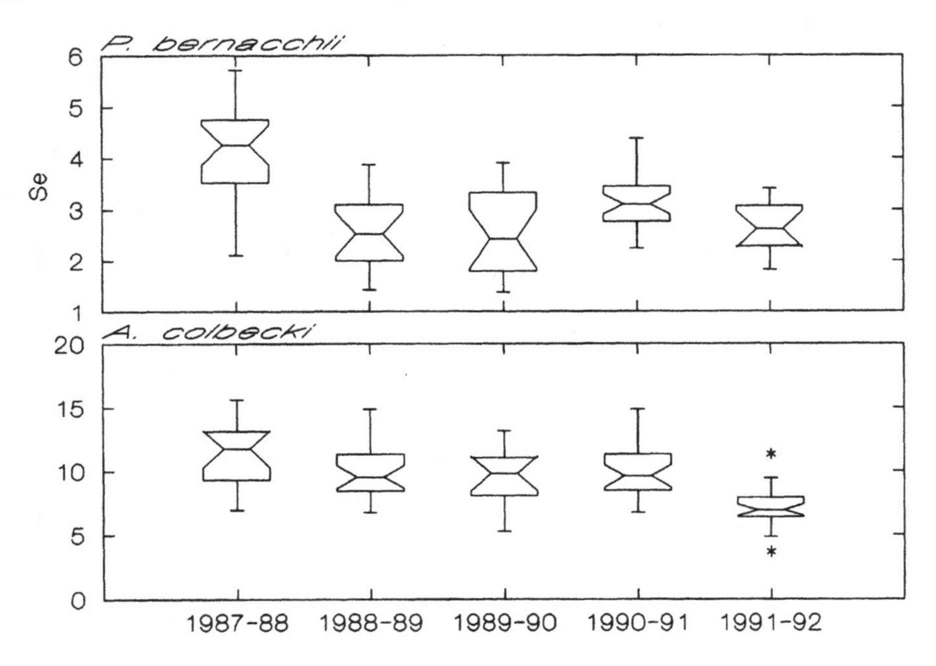

Figure 4 Comparison of selenium concentrations, expressed as µg/g dry weight, in muscular tissue of *P. bernacchii* (top), and in soft parts of *A. colbecki* (bottom), observed from 1987 to 1992. Outliers have been plotted as * (> 1.5 times the interquartile range).

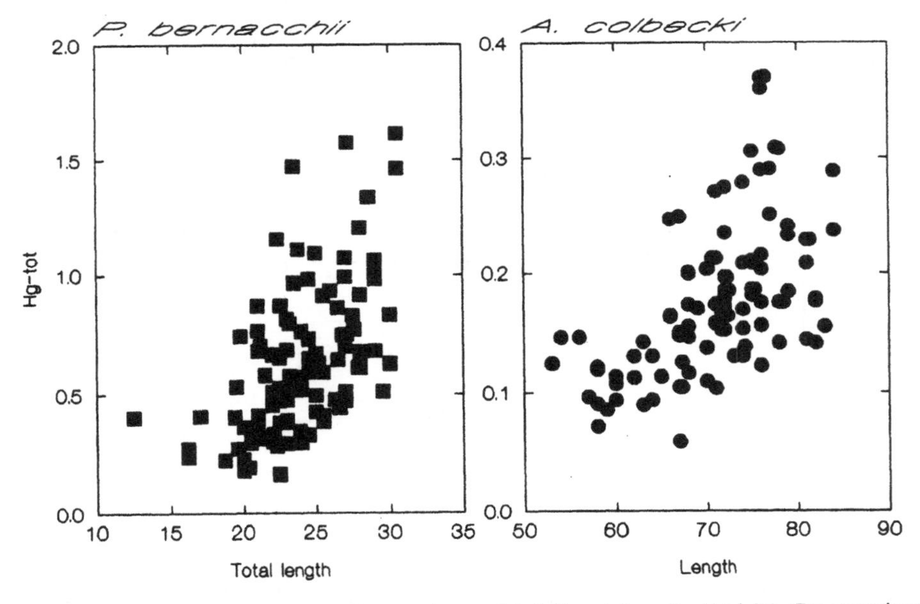

Figure 5 Mercury concentration vs. size in *P. bernacchii* (left) and *A. colbecki* (right). Concentrations are expressed in µg/g dry weight, size for *P. bernacchii* in cm, and for *A. colbecki* in mm.

Selenium concentrations are higher in *A. colbecki*, and this is consistent with what was found by Thibaud and Noel[17] in the Mediterranean for two species (*Mytilus galloprovincialis* and *Sparus aurata*) of analogous trophic levels. Again, the two species differ in variations, *P. bernacchii* showing the larger difference in concentrations.

CONCLUSIONS

The annual differences observed in the concentrations of the trace elements considered, were not the same for the two species, thus making difficult to identify the cause for such variations.

Linear trend analysis for temporal variations was done on the variables measured, using the median values for each year. For *P. bernachii* the percentage of organic mercury shows a positive trend ($r = 0.935$), while for *A. colbecki* the selenium decreases with the time ($r = -0.882$). Obviously more data are needed to support the trends found, and research should be extended for several consecutive years, if any variation for trace elements in the environment have to be detected.

Data available in literature[18,19], for total mercury, concerning blue mussel (*Mytilus edulis*), flounder (*Platichthyus flesus*), and plaice (*Pleuronectes platessa*) collected in areas of the Northern emisphere far from any point source of contamination, show that no evident temporal trend can be detected.

The results obtained, despite the large number of specimens analysed, probably reflect variations in the biological conditions of the organisms rather than the environmental concentrations.

Data reported form, therefore, the basis for the knowledge of existing levels for mercury and selenium in two species widely distributed in the Antarctic seas. Medians obtained during the five years in *P. bernacchii* are: 0.60 μg/g dry weight for total mercury, 78% for the percentage of organic mercury, and 3.0 μg/g dry weight for selenium. Values found in *A. colbecki* are: 0.17 μg/g dry weight for total mercury, 28% for the percentage of organic mercury, and 9.1 μg/g dry weight for selenium.

Acknowledgements

This work was funded in the framework of the Italian National Program of Researches in Antarctica (PNRA).

References

1. W. L. Stockton, *Mar. Biol.*, **78**, 171–178 (1984).
2. I. Di Geronimo and A. Rosso, *Nat. Sc. Com. Ant., Ocean. Camp. 1987–88, Data Rep.*, **I**, 407–421 (1990).
3. M. Vacchi, A. Castelli, M. La Mesa and M. Taviani, *2nd Int. Conf. 1. Biology of Antarctic Fishes, Ravello (Italy), May 30–June 1. Conf. Abstr.*, 86–87 (1990).
4. M. Vacchi, S. Greco and M. La Mesa, *Memorie di Biologia Marina e di Ocenografia*, **19**, 197–201 (1991).
5. D. J. H. Phillips, *Environ. Pollut.*, **13**, 281–317 (1977).
6. W. Fischer and J. C. Hureau (eds.), 1985. FAO species identification sheets for fishery purposes. Southern Ocean (Fishing areas 48, 58 and 88) (CCAMLR Conservation Area). Prepared and published with the support of the Commision for the Conservation of Antarctic Marine Living Resources, Rome, FAO, Vol 1: 1–232, Vol 2: 233–470.
7. R. Capelli, R. De Pellegrini, V. Minganti and E. Amato, *Annali di Chimica*, **79**, 561–569 (1989).
8. R. Capelli, V. Minganti, R. De Pellegrini and F. Fiorentino, *Environmental Impact in Antarctica*, Roma, 8–9 June 1990, 47–54.

9. V. Minganti, F. Fiorentino, R. De Pellegrini and R. Capelli, *Intern. J. Environ. Anal. Chem.*, **55**, 197–202 (1994).
10. R. Capelli, V. Minganti, F. Fiorentino and R. De Pellegrini, *Annali di Chimica*, **81**, 357–369 (1991).
11. J. M. Thompson, *Analytical Proceedings*, **25**, 363–365 (1988).
12. M. D. Nicholson and R. J. Fryer, *Mar. Pollut. Bull.*, **24**, 146–149 (1992).
13. U. Föster and G. T. W. Wittman, *Metal Pollution in the Aquatic Environment.* Second Revised Edition. Springer-Verlag, Berlin-Heidelberg-New York, pp. 486 (1981).
14. M. Bernhard, *UNEP Regional Seas Reports and Studies No. 98*, UNEP, 1988.
15. C. Amiard-Triquet, *Bull. Ecol.*, **20**, 129–151 (1989).
16. R. Gächter and W. Geiger, *Schweiz. Z. Hydrol.*, **41**, 277–290 (1979).
17. Y. Thibaud and J. Noel, in: *UNEP/FAO, Final Reports on research projects dealing with mercury, toxicity and analytical techniques. MAP Technical Reports Series, No. 51. UNEP*, Athens, (1991) 1–18.
18. J. Stronkhorst, *Mar. Pollut. Bull.*, **24**, 250–258 (1992).
19. L. A. Jørgensen and B. Pedersen, *Mar. Pollut. Bull.*, **28**, 235–243 (1994).

EVOLUTION OF CADMIUM AND LEAD CONTENTS IN ANTARCTIC COASTAL SEAWATER DURING THE AUSTRAL SUMMER

G. SCARPONI[a,*], G. CAPODAGLIO[a,b], C. TURETTA[a], C. BARBANTE[a,b], M. CECCHINI[a], G. TOSCANO[a] and P. CESCON[a,b]

[a]*Department of Environmental Sciences, University of Venice Ca' Foscari, I-30123 Venice, Italy;* [b]*Centre for Studies on Environmental Chemistry and Technology-CNR, University of Venice Ca' Foscari, I-30123 Venice, Italy*

The seasonal evolution of the cadmium and lead distribution in the water column of the Gerlache Inlet (Ross Sea) was studied during the 1990–91 austral summer. Measurements were carried out by Anodic Stripping Voltammetry in Antarctica immediately after the collection and filtration of samples. The concentrations of both metals were homogeneous before the phytoplankton bloom with mean values of 0.71 (SD 0.10) and 0.116 (SD 0.014) nmol/l for cadmium and lead respectively. A subsequent depletion in metal concentration was observed in the shallow waters. The surface concentration of cadmium decreased to about 0.1 nmol/l at the end of the season. The vertical distribution of lead was less affected by the seasonal evolution and the mean surface concentration decreased to 0.044 nmol/l in the same period. The results are evaluated with respect to physical and biological processes in the area examined and compared with those obtained on previous expeditions in the same area.

Keywords: Cadmium; lead; DPASV; seawater; Terra Nova Bay; Antarctica; seasonal evolution

INTRODUCTION

The coastal seas of the Antarctic Continent are subjected to meteorological, climatic and environmental conditions which differentiate them markedly from all other marine ecosystems. As well as the extreme climatic conditions and

*Corresponding author. Fax: +39-41-5298549.

the remoteness from all direct pollution sources, seasonal change is reduced to two long periods. The winter is characterised by darkness and the presence of sea ice; in this period photosynthesis is precluded and only the respiration and decomposition of organic compounds are possible. The summer is characterised by a continually light period, the dissolution of pack ice, a sudden phytoplankton bloom and, at the end of the summer, the formation of new marine ice.

In the framework of the Italian Research Programme in Antarctica, investigations were carried out into the distribution and speciation of some trace metals (Pb, Cd, Cu) in the surface seawaters of Terra Nova Bay (Ross Sea) and the base levels of metal concentrations during the austral summer were established in three successive expeditions (1987–88, 1988–89 and 1989–90)[1–6]. The metal concentrations were in good agreement with values obtained by other researchers in the Southern Ocean[7–13], although data variability was higher than that observed in the Weddell Sea and the Indian sector of the Southern Ocean[7–10]. In particular our previous studies on cadmium concentration showed high variability both spatially and temporally[2,5] and a depletion during the summer[2].

However, our previous investigations were carried out on sub-surface samples; no information was available for the vertical distribution and nor was the complete temporal evolution investigated.

During the 1990–91 expedition a more systematic study of the summer seasonal evolution of metal concentration in the water column of the Gerlache Inlet (in Terra Nova Bay) was carried out in order to explain the variability of the previously observed data and to highlight possible anthropogenic contamination. The investigation began well before the dissolution of pack ice and carried on until the end of the austral summer. Samples were analysed on site by one of the authors (G.S.) immediately after sampling in the clean chemistry laboratory of the Italian Station at Terra Nova Bay.

In this paper we present the results obtained on samples collected during the 1990–91 expedition for cadmium and lead, two elements characterised by different geochemical properties. Cadmium, although it is not known to be essential biologically, exhibits a nutrient-type distribution (recycled element) and is involved in shallow regeneration cycles similar to those of phosphate and nitrate[14]. The influence of processes of production and remineralization of biogenic particles on the distribution of dissolved cadmium has already been reported in the literature for different geographical areas[15,16]. Lead is a typical scavenged element substantially influenced by local inputs[17]. Results of the present study will be compared with those obtained in the same area from previous expeditions.

EXPERIMENTAL

Sampling

During the 1990–91 austral summer seawater samples were collected along a water column of the Gerlache Inlet (called station B; 74°40′07″ S, 164°07′15″ E; distance from the Italian Station of Terra Nova Bay about 5 km; sea floor about 270 m) on Nov. 29, Dec. 7 and 26, Jan. 6 and 30, and Feb. 11 (see Figure 1). During this period the site was covered by pack until about mid-January. In general samples were collected at the following depths (in m): 0.5, 10, 25, 50, 100, 250. On Nov. 29 and Dec. 7 seawater was gathered only from 0.5, 25 and 100 m depths. On February 11th the site selected for the study was occupied by an iceberg and samples were collected a few hundred metres away from it (74°40′02″ S, 164°07′45″ E); in this case the bottom sample was drawn at a depth of 290 m instead of the usual 250 m due to the deeper sea floor (315 m).

In the presence of pack, samples were collected through a hole drilled by an ice corer (Duncan, UK, Model BTC). Subsurface water (about 0.5 m below the seawater level) was collected by an air-driven diaphragm pump in Teflon (Disco, Italy, Mod. DL15) and temporarily stored in a 50 1 polyethylene tank until filtration (see below). Deeper samples were collected by 20 or 30 1 Go-Flo samplers (General Oceanics, USA), coated with Teflon internally and with pressurization capability. A non-metallic (Kevlar) hydrowire and a Teflon covered ballast attached to the wire at least 10 m below the sampler, were used.

Samples were filtered through membrane filters of 0.45 μm pore size immediately after sampling (within 2–3 hours), by pressurizing (pressure lower than 0.5 bar) the polyethylene tank or the Go-Flo bottle with pure nitrogen (chromatographic grade) and pushing the sample through a Teflon filtration apparatus (Sartorius, holder for in-line pressure filtration, SM 16540; cellulose nitrate membrane filter, SM 11306, 142 mm diameter). After discarding a first aliquot of about 3–5 litres of water, filtered samples were collected directly in 2 1 FEP bottles and stored at +4°C until analysis (carried out on site within 1–2 weeks). Aliquots of samples were also stored frozen (−20°C) for speciation analysis to be performed later in Italy (not considered here).

Polyethylene tanks, Go-Flo bottles, filtration apparatus, filters and storage bottles were acid cleaned and conditioned to minimise sample contamination from the material used; the detailed cleaning procedures have been reported previously[18]. Handling and treatment of the samples were carried out in the clean chemistry laboratory (Class 100) available at the Terra Nova Bay Station in Antarctica.

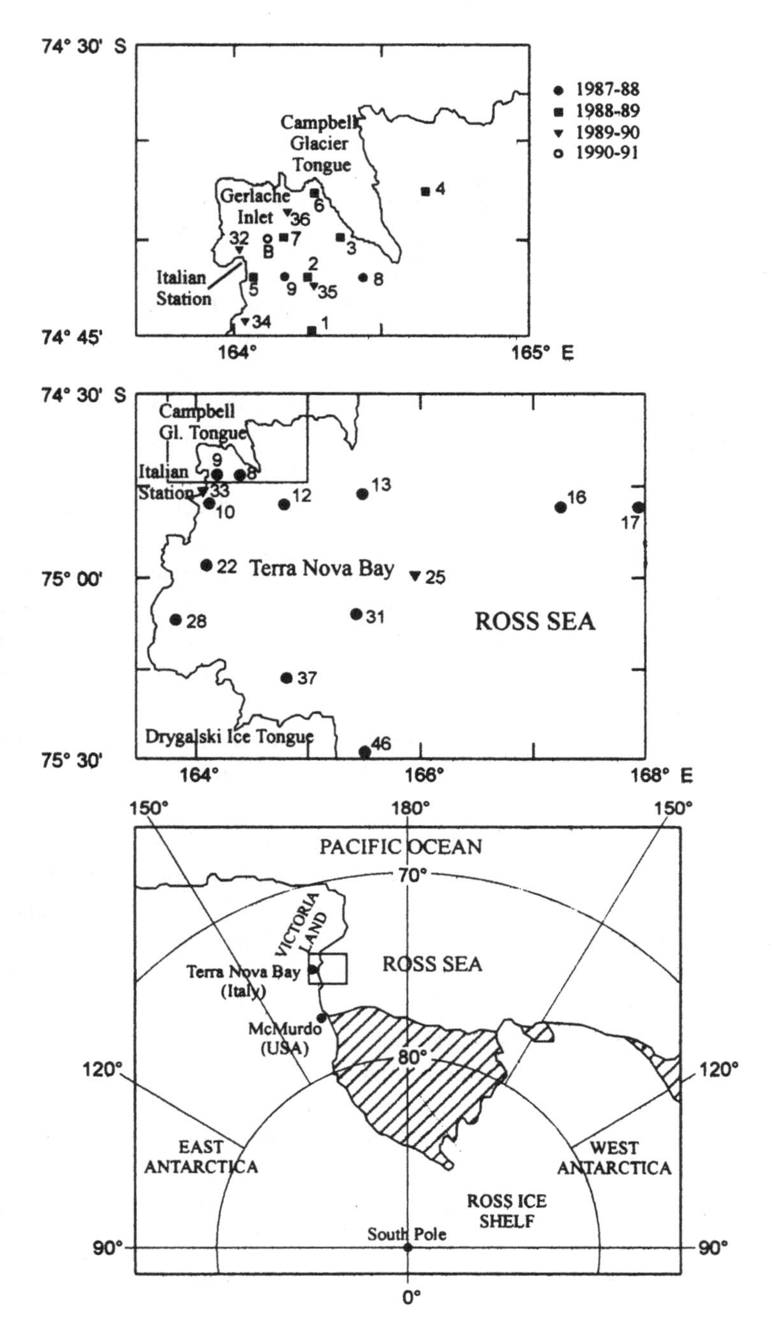

FIGURE 1 Sampling stations in Terra Nova Bay during expeditions from 1987–88 to 1990–91.

Subsurface samples collected during previous expeditions (1987–88: sample ref. 8, 9, 10, 12, 13, 16, 17, 22, 28, 31, 37 and 46; 1988–89: ref. 1, 2, 3, 4, 5, 6 and 7, in positions 5, 6, 7 samples were collected before (ref. 5P, 6P, 7P) and after pack melting; 1989–90: ref. 25, 32, 33, 34, 35 and 36) were collected and treated in similar way. Details on procedures and locations (see also Figure 1) have already been reported[1,2,4].

The sampling/filtration procedure was previously tested to be contamination-free through an international intercalibration exercise on seawater sampling for trace metal determination[19,20].

Temperature (T), salinity (S) and chlorophyll concentration (Chl) were measured along the water column by a multiparametric probe (ME, Meerestechnik), equipped with a back scattering fluorometer (Dr. Haardt, Backscat). Unfortunately the probe was available on-site for the present study only for a limited period, between December 15 and January 15. Thus T, S and Chl profiles are available only for observations carried out on December 26 and January 6.

Voltammetric Analysis

Metal concentrations were determined by Differential Pulse Anodic Stripping Voltammetry (DPASV) carried out on a Thin Mercury Film Electrode (TMFE). An EG&G 384B Polarographic Analyzer was used. The electrochemical cell (EG&G, mod Rotel 2), specifically designed to detect ultra-low metal concentrations present in sea water, has been described previously[18,21]. The cell is equipped with a Rotating Glassy Carbon Disc Electrode (RGCDE), on which the TMFE is electrochemically plated, an Ag/AgCl,KCl(sat.) reference electrode and a platinum auxiliary electrode.

The mercury film was prepared daily immediately prior to each analysis using the procedure summarized below[1,2,4,18]. The graphite surface was polished using wetted alumina (0.05 μm grain size, 1000 rpm rotation rate), then washed with diluted (1:200) ultrapure HCl (Merck, Suprapur) and plentifully rinsed with ultrapure water (Millipore, Milli-Q). The mercury deposition was obtained by electrolysis from a solution containing KCl about 2.5×10^{-2} mol/l and $Hg(NO_3)_2$ 10^{-4} mol/l at -1.000 V for 20 min using a rotation rate of 4000 rpm. After the film formation the cell cleanup was tested by a differential pulse voltammetric scan in the positive direction; if the blank voltammogram did not show peaks and the base line was satisfactory analysis of the sample (previously outgassed for 20 min with nitrogen flow) was started. Before transferring the sample to the measurement position the electrochemical cell was rinsed with ultrapure water previously outgassed with ultrapure nitrogen (chromatographic grade).

Total dissolved metal determinations were carried out on samples digested in ultrapure HCl for at least 24 hours. Digestion was carried out directly into the Teflon vessel used as the cup of the electrochemical cell adding 100 µl of HCl (32%, NIST, Gaithersburg MD, U.S.A.) to approximately 50 ml of the sample (pH about 2). A more precise measurement of the volume was carried out subsequently, at the end of the determination (see below). Tests carried out to verify the efficiency of digestion procedure confirmed the complete release of the metals from organic ligands[1,2].

Voltammetric measurements were performed applying the following experimental conditions: deposition potential −0.950 V, deposition time 20 min, electrode rotation rate 4000 rpm, equilibration period 30 s, mode DPASV, scan rate 10 mV/s, pulse amplitude 50 mV, pulse frequency 5 s^{-1}, final potential −0.180 V. After the voltammetric scan the electrode was allowed to remain at a potential of −0.200 V for 5 min to ensure complete removal of the amalgamated metals. Quantification was obtained by the multiple standard additions method; three spikes of metal standard solutions were always made. At the end of the measurement cycle the sample volume was detected with more precision (±0.5 ml) using a cylinder.

The blank for the digestion procedure was obtained by adding the ultrapure HCl to a KCl solution (about 3×10^{-2} mol/l prepared with ultrapure water and KCl of Suprapur grade, Merck) and analysing it before and after the acid treatment. The concentration difference was always lower than the detection limit for Cd and Pb (about 0.2 ng/l) so the blank contribution was considered to exert no influence on the determination.

Statistical Analysis

A multivariate statistical approach was applied to analyse data of the 1987–88 and 1989–90 seasons. Samples were characterized through measurement of the concentration of metals[1,2,4–6], nutrients[22,23], chlorophyll[24–26], temperature[24,27,28] and salinity[23,25,28]. Details of the variables used are given below. Unfortunately the 1990–91 season cannot be included in the present comparison because of the lack of data concerning other than trace metals.

Principal component analysis (PCA)[29] was applied to identify variable associations, to obtain a dimension reduction and to observe the group structure of the data. PCA was carried out on standardized data due to the different measurement units, magnitude and variability of the variables.

The UNISTAT[30] statistical package for computer was used for all the required computations.

HYDROGRAPHY

The Weddell and Ross seas are the areas where the larger part of the Antarctic Bottom Water is formed. Because of its high density the Bottom Water flows down the continental slope into the South Atlantic and eastward through the Indian and Pacific Ocean sectors of the Southern Ocean[31].

The hydrological characteristics of the Ross Sea and Terra Nova Bay (a small bay west of the Ross Sea, between Cape Washington and the Drygalski Ice Tongue) were studied during earlier oceanographic cruises and a general circulation pattern of water masses has been proposed[22–24,27,28,32–36]. The Ross Sea is affected by a branch of the Antarctic Coastal Current which, with a cyclonic movement, follows the coast from east to west and successively from south to north through the McMurdo Sound and Victoria Land[31,33,35]. However, the coastal irregularities, bottom topography and localized phenomena, such as polynyas, give rise to several deviations from this general circulation, principally on a small scale. During the summer the surface layer features a low salinity (<34.5‰) typical of Antarctic Surface Water (AASW), whose properties are affected by processes such as seasonal warming or surface freezing and interaction with meltwater from continental or sea ice; in particular, higher temperatures are observed in the southern part of the Sea where pack opens up earlier in the season[32,33]. Circulation below the AASW is controlled by Circumpolar Deep Water (CDW), characterized by relatively high temperature and salinity. This water type represents the only water of external origin near the continental shelf, so it gives rise to the other water types by mixing and cooling and by input of precipitation, meltwater or brine. On the basis of salinity and $\delta^{18}O$ measurements it has been proposed that the combination of brine deriving from the freezing of sea surface, glacial meltwater and precipitation could generate the High Salinity Shelf Water (HSSW), principally in the Western Ross Sea although the formation mechanism has not been completely clarified[33]. Flow of CDW along the continental slope helps to confine the HSSW in the depression of the Shelf, particularly, along the shore of Victoria Land[32,33].

In Terra Nova Bay the structure of the water column in summer is quite simple. It is generally characterised by a relatively warm and low salinity surface layer of 10–50 m ascribed to AASW over a homogeneous saltier and colder water extending down to the floor and tending to HSSW[24,27,28,35,36]. The physical and chemical characteristics of surface water, which presents variable currents principally driven by the wind, are affected by local processes and seasonal evolution. Significant interannual variability was observed for the salinity and temperature of the surface layer during the 1987–88 and 1989–90 seasons (average values were 1.0 °C and 34.19 ‰ in 1987–88[27] and 0.5 °C and 33.58 ‰ in

1989–90[14,28]). Deep waters, as normally observed, are much more homogeneous and reproducible, with temperature and salinity of about −1.9 °C and 34.8 ‰, respectively[24,27,28], i.e. characteristic values of Ross Sea HSSW[33]. The hydrodynamics of deep layers show more stable currents and a low vertical component[35], especially in coastal areas. The latter circumstance could be explained, as supposed by Stocchino and Lusetti[36] by the presence of a relative temperature minimum delimiting the surface layer.

In the Gerlache Inlet the summer current shows a clockwise direction both in the surface and in the deeper layers[35]. Temperature and salinity profiles measured during the pack melting in each case show a warm surface layer and a deep water with characteristics tending to HSSW (see Figures 2a and 3a). The T versus S plot (see Figures 2b and 3b) enables the water masses observed in the Gerlache Inlet to be characterized on the basis of domains identified in the Ross Sea[33,37]. The surface presents a very thin, well stratified layer (a few meters) having low temperature (−1 °C) and low salinity (<34 ‰), probably

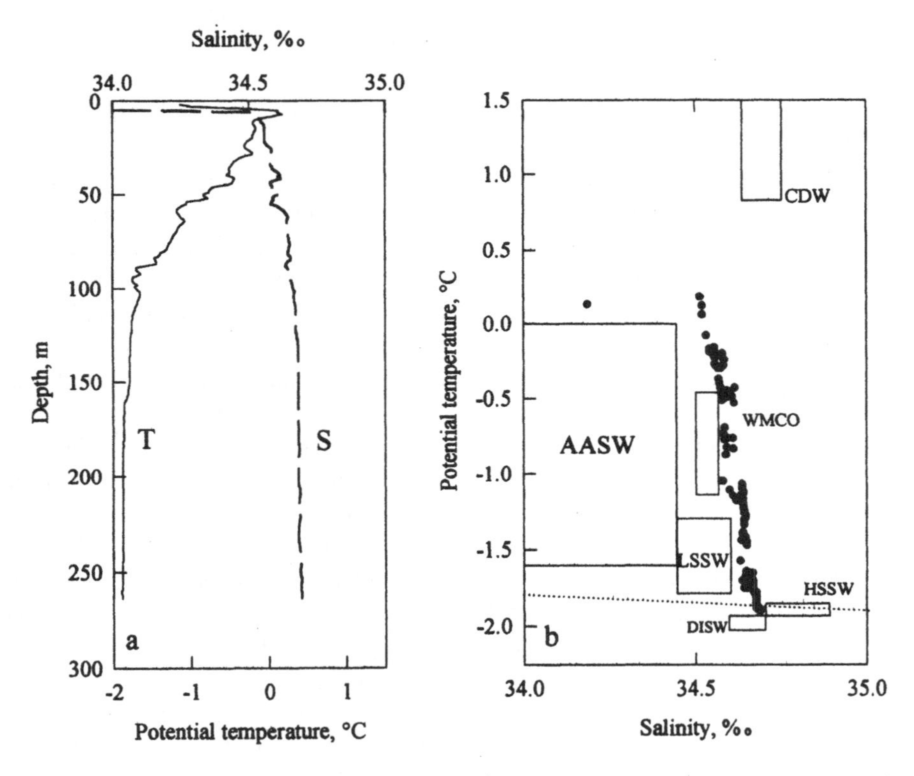

FIGURE 2 Vertical profiles for salinity and potential temperature (a), and potential temperature/salinity plot (b) for December 26, 1990. Abbreviations and domains defined in Jacobs et al.[33].

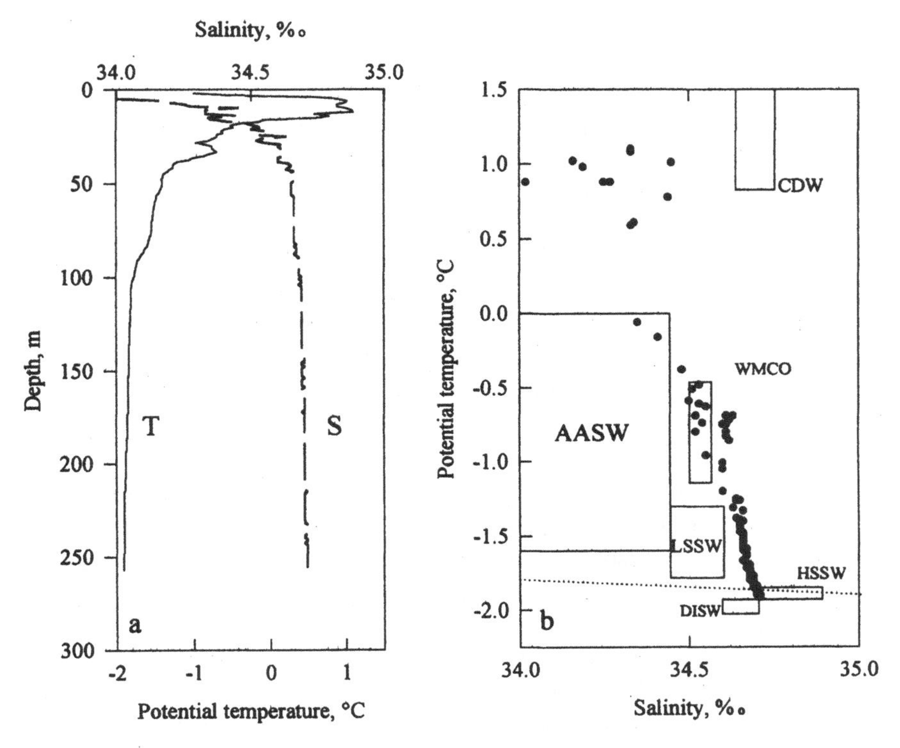

FIGURE 3 Vertical profiles for salinity and potential temperature (a), and potential temperature/salinity plot (b) for January 6, 1991. Abbreviations and domains defined in Jacobs et al.[33].

resulting from recent pack melting. Beneath it are two major water masses. An upper layer of 20–100 m can be distinguished from the deep waters; this layer shows quite different characteristics for the two profiles. On December 26 the first 20 m layer assumes a temperature of about 0 °C and lies on a layer (comprised between 20 and 100 m) which shows a gradual transition to HSSW (temperature between 0 and −1.9 °C, salinity between 34.5 and 34.7 ‰). This latter water mass, presenting characteristics similar to a warm core (WMCO) (see Figure 2b) could represent the result of intrusion of water masses related to the winter polynya in the Terra Nova Bay[33,35]. On January 6, immediately before pack melting, the upper layer presents a temperature of about 1 °C and a salinity of about 34.3 ‰, showing characteristics similar to AASW (see Figure 3b), although the temperature assumes higher values than those reported by Jacobs et al.[33]. Below this upper layer, a water mass can be recognized with characteristics similar to WMCO (which presents a relative temperature minimum at a depth of 30 m) and an intermediate homogeneous layer of between

50 and 80 m with a mean temperature of −1.5 °C and a salinity of 34.66 ‰; deeper, the water mass shows a gradual transition to HSSW. Measurements are not available for the complete seasonal evolution, so it is impossible to evaluate if the hydrodynamic regime was different before December 26 and after January 6, but the available profiles seem to show the intrusion in the top layers of Gerlache Inlet of surface water from the Terra Nova Bay during the pack melting.

Distribution of nutrients in Terra Nova Bay during the seasons 1987–88 and 1989–90 were reported by Catalano et al.[22,23]. Vertical profiles showed a surface depletion, affecting the first 20–50 m layer, where the concentrations were about halved with respect to deeper layers. The concentrations reported for deep water are in excellent agreement with values observed in the Weddell Sea where, contrary to Terra Nova Bay, only slight surfacial depletion was observed[7,9].

Considering for example the 1989–90 data set, and particularly the coastal subset (more relevant for the present study) a few interesting observations can be made. Nitrate, phosphate and silicate concentrations were well correlated and, as usual, correlation coefficients increased in passing from data referred to the upper 25 m surface layer ($0.89 \leq$ corr ≤ 0.96) to data of complete vertical profiles (all correlations equal to 0.98). Moreover the nitrate/phosphate molar concentration ratio, for the complete profiles, was 14, a value in excellent agreement with typical values reported for the Antarctic ocean[7], and slightly lower than that estimated for global oceanic waters, i.e. 16, by Redfield et al.[38].

As regards the relationships between nutrients and salinity contrasting results can be observed if one considers shallow and deep samples separately. In particular the correlation of nitrate, phosphate and silicate with salinity is significantly lower in shallow (corr 0.64, 0.67 and 0.69, respectively) than in deep waters (corr 0.97, 0.95 and 0.87). Moreover, the slope of the regression line obtained by plotting nitrate as a function of salinity was 6.6 μM/psu for shallow samples and 21.6 μM/psu for deep samples.

In conclusion, the presence of a marked depletion of nutrients in the shallow layer of the coastal area of Terra Nova Bay, together with nutrients/salinity relationships which show correlation and regression coefficients which are significantly lower than those obtained in deep waters, emphasize an inefficient regeneration of nutrients in the upper part of the water column, which seems in agreement with the low vertical flux measured by Stocchino and Manzella[35]. As a consequence, it can be concluded that the nutrient distribution is mainly influenced by physical processes in deep waters and by biological activity in the shallow waters.

RESULTS

Results of cadmium and lead concentration measurements as a function of the depth for the samples collected during the 1990–91 expedition are reported in Tables I and II. Two or more replicates were carried out for all the samples except for those referring to the last sampling because very little time was available before the closure of the field laboratory.

Table III reports data related to surface samples collected during the 1987–88 and 1989–90 expeditions, to be considered for the statistical analysis. Cadmium and lead concentration, salinity, temperature, chlorophyll and nutrients content are given. References are cited therein.

Cadmium

The distribution of cadmium in the water column (see Figure 4) presents different features according to the period of observation.

At the beginning of the summer season, until about mid-December (and presumably during the winter), the vertical profile of Cd concentration is homogeneous with a mean value of 0.71 (SD 0.10) nmol/l. The relatively high concen-

TABLE I Cadmium concentration in samples collected in Gerlache Inlet during the 1990–91 Italian expedition. In parentheses mean values.

Depth m	*Cadmium concentration, nmol/l*					
	Nov 29	*Dec 7*	*Dec 26*	*Jan 6*	*Jan 30*	*Feb 11*
0.5	0.76, 0.68, 0.79, 0.72, 0.72 (0.73)	0.86, 0.85, 0.97, 0.93 (0.90)	0.72, 0.71, 0.66, 0.83 (0.73)	0.19, 0.19 (0.19)	0.17, 0.15, 0.15 (0.16)	0.10 (0.10)
10			0.10, 0.10, 0.11, 0.13 (0.11)	0.14, 0.15, 0.18 (0.16)	0.18, 0.22 (0.20)	0.10 (0.10)
25	0.61 (0.61)	0.70, 0.69 (0.70)	0.14, 0.10, 0.10 (0.11)	0.17, 0.14 (0.16)	0.26, 0.25, 0.26 (0.26)	0.11 (0.11)
50			0.26, 0.32, 0.26 (0.28)	0.61, 0.55, 0.57 (0.58)	0.30, 0.32, 0.31 (0.31)	0.12 (0.12)
100	0.67, 0.67 (0.67)	0.67, 0.64, 0.67, 0.66 (0.66)	0.64, 0.70, 0.68 (0.67)	0.61, 0.69 (0.65)	0.63, 0.74 (0.68)	0.54 (0.54)
250			0.76, 0.66, 0.55 (0.66)	0.64, 0.75, 0.75, 0.80, 0.68 (0.72)	0.34, 0.31, 0.37 (0.34)	
290						0.25, 0.28 (0.26)

TABLE II Lead concentration in samples collected in Gerlache Inlet during the 1990–91 Italian expedition. In parentheses mean values

Depth, m	*Lead concentration, nmol/l*					
	Nov 29	*Dec 7*	*Dec 26*	*Jan 6*	*Jan 30*	*Feb 11*
0.5	0.102, 0.100, 0.087, 0.085, 0.108 (0.096)	0.106, 0.110 (0.108)	0.102, 0.100, 0.119, 0.094, 0.114 (0.106)	0.056, 0.074 (0.065)	0.068, 0.042, 0.060 (0.057)	0.044 (0.044)
10			0.088, 0.070 (0.079)	0.030, 0.034, 0.044 (0.036)	0.027, 0.025 (0.026)	0.037 (0.037)
25	0.114 (0.114)	0.117, 0.142 (0.130)	0.030, 0.040, 0.044 (0.038)	0.261, 0.218 (0.240)[a]	0.027, 0.023 (0.025)	0.028 (0.028)
50			0.109, 0.138, 0.124 (0.124)	0.052, 0.040, 0.044 (0.045)	0.043, 0.049 (0.046)	0.034 (0.034)
100	0.134, 0.130 (0.132)	0.109, 0.119, 0.127 (0.118)	0.080, 0.118, 0.116 (0.105)	0.133, 0.123 (0.128)	0.024, 0.025 (0.024)	0.053 (0.053)
250			0.130, 0.093, 0.113 (0.112)	0.123, 0.119, 0.096, 0.133, 0.102 (0.115)	0.139, 0.144, 0.131 (0.138)	
290						0.116 (0.116)

[a]Contaminated sample.

TABLE III Metal concentrations[1,2,4–6] and other hydrological data[22–28] obtained in the surface water of Terra Nova Bay in the 1987–88 and 1989–90 seasons

Station	Cd nmol/l	Pb nmol/l	Cu nmol/l	Salinity ‰	Chlorphyll mg/l	Temperature °C	SiO_4^{4-} µmol/l	PO_4^{3-} µmol/l	NO_3^- µmol/l	NO_2^- µmol/l
					1987–88 Campaign					
8	0.24	0.063	4.3	34.85	0.83	1.99	56.9	0.68	15.6	0.06
9	0.19	0.114	2.8	34.61	0.44	2.10	64.9	0.79	16.9	0.04
10	0.26	0.061	2.4	34.82	1.20	2.00	61.3	0.64	15.2	0.04
12	0.15	0.035	1.6	34.81	0.76	0.80	61.3	0.93	22.3	0.05
13	0.19	0.032	1.4	34.84	1.04	0.74	54.6	0.84	20.3	0.05
16	0.08	0.026	1.8	34.10	1.10	1.02	52.4	0.79	18.4	0.08
17	0.10	0.038	0.9	32.80	1.44	0.03	54.2	0.81	20.1	0.25
22	0.09	0.038	1.4	34.75	0.85	0.60	65.6	1.14	20.3	0.04
28	0.13	0.034	0.9	34.69	0.36	1.65	63.6	1.05	17.5	0.04
31	0.14	0.025	2.0	34.52	2.44	0.86	53.4	0.79	18.4	0.07
37	0.15	0.027	1.6	34.33	1.83	0.73	57.5	0.79	18.0	0.10
46	0.40	0.029	2.7	33.78	2.35	−0.23	55.3	1.01	19.2	0.14
					1989–90 Campaign					
25[a]	0.071	0.034	1.9	34.50	4.10	−0.75	68.0	1.46	24.6	0.15
32	0.094	0.032	1.7	33.72	2.82	0.77	18.5	0.62	8.8	0.28
33	0.053	0.015	2.1	33.70	1.44	1.28	15.3	0.31	9.6	0.25
34	0.036	0.031	2.5	33.81	2.48	0.19	24.6	0.83	11.5	0.22
35	0.046	0.032	1.6	33.25	0.18	0.85	7.6	0.32	6.2	0.32
36	0.027	0.018	—[b]	33.62	3.68	0.84	16.0	0.47	8.5	0.24

[a]Depth 20 m. [b]In PCA the missing value was replaced by the mean value of the variable for the 1989–90 data set.

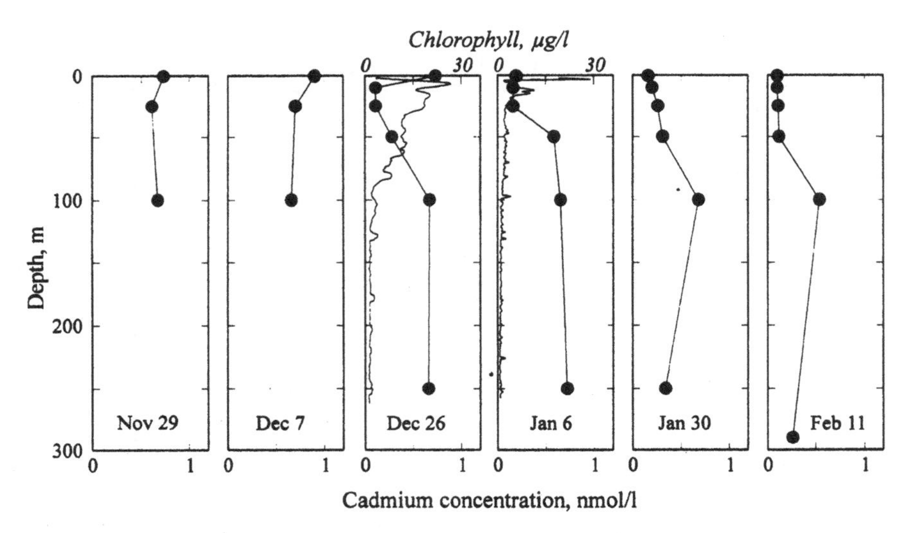

FIGURE 4 Seasonal evolution of depth profile for cadmium concentration (•) and chlorophyll (—). 1990–91 campaign.

tration level and the homogeneity are accounted for by the upwelling of enriched deep water to the surface and by the absence of an efficient scavenging process or uptake by organisms. Values are in agreement with typical concentrations observed in deep waters of Antarctic regions[7,9] (Weddell Sea, 0.5–0.9 nmol/l) and Subantarctic regions[16] (Pacific sector, 0.58–0.65 nmol/l), and similar homogeneous profiles are observed in upwelling areas of the Weddell Sea[7,9]. Lower values are reported for Antarctic seawater in samples collected later in the summer season by us[2,5], 0.026–0.69 nmol/l Ross Sea, and by others[11], 0.15–0.48 nmol kg^{-1} Weddell Sea, but higher results are also reported for the Atlantic-Indian sector of the Southern Ocean[12], 0.91–1.28 nmol kg^{-1}

Later in the season (see Table I), contemporarily with ice melting and with the development of the phytoplankton bloom (see profiles of December 26 and January 6), a marked depletion of the cadmium concentration is observed. The depletion, not present on December 7 and already underway on December 26 (when a minimum of 0.11 nmol/l was observed at a depth of 10–25 m), extends gradually to a depth of 50 m after the pack ice is completely melted (see profile of January 30). Finally (February 11) the entire 50-m upper layer presents an almost homogeneous concentration with a mean value of 0.11 (SD 0.01) nmol/l.

The last two profiles present maxima at a depth of about 100 m and unexpected lower values in the bottom. At present no explanation is given for this apparently anomalous behaviour which is nevertheless observed in two profiles;

a more complete interpretation of these data would require a more detailed knowledge of local marine currents.

The observed depletion is much more pronounced than that shown in profiles obtained in regions characterized by an efficient upwelling of the Weddell Sea[7,9] for which, however, the temporal variation (behaviour) is not available. Conversely an analogous remarkable depletion is reported for the surface layers of the Scotia Sea[7] and of the Subantarctic region southwest of New Zealand[16].

Two hypotheses can be formulated to explain the quite rapid depletion of Cd concentration in shallow waters of the Gerlache Inlet as observed in the temporal investigation carried out during the summer 1990–91.

The first explains the phenomenon in terms of merely the physical effect of marine currents. It considers that a surfacial intrusion of low salinity, low Cd content water mass (AASW) enters the Gerlache Inlet from the Bay, interrupts the (slowly flowing) upwelling and extends rapidly to the first 50-m layer. In this event, because the depletion would derive from a dilution of surface waters, the normal correlation of Cd concentration with salinity should still be retained in the surface layer.

According to the second hypothesis, Cd depletion has a biological origin, due to interaction of metal with particles produced by processes taking place during the phytoplankton bloom; the Cd distribution would then follow that of major nutrients. In this event cadmium/salinity and major nutrients/salinity plots would show, for shallow waters, a different behaviour with respect to the lower water column and, consequently, lower correlation should be obtained for the entire data set.

To gain more insight into the cause of surface cadmium depletion it is useful to consider the relationships of the metal concentration with temperature, chlorophyll and salinity for the two profiles for which corresponding data are available (i.e. December 26 and January 6). Here clear negative correlations are observed for cadmium concentration with temperature (corr -0.82) and chlorophyll (corr -0.71) (see also Figure 4). Conversely, no correlation is observed between Cd concentration and salinity (corr 0.07), although high correlation emerges when one considers only samples from deeper than 25 m (corr 0.75). The latter evidence together with: (i) the available chlorophyll profiles (Figure 4), which show the highest concentrations in the 50–100 m surface layer, (ii) the above observations on nutrient distribution and on the low vertical fluxes in coastal area of Terra Nova Bay, and (iii) the general knowledge about the behaviour of cadmium as a nutrient-type element[17], suggest that an important explanatory factor for the cadmium depletion is the quick summer increase in biological activity in the whole area. Of course the contribution of the transport of depleted water mass from the centre of the bay (where the biological processes may have begun) through the existing current to the Inlet, particularly below the pack ice in the early summer cannot be excluded; indeed this could partly explain the rapidity of the phenomenon.

In conclusion, the joint effect of biological and physical processes can very probably explain the rapid cadmium depletion observed in the Gerlache Inlet within 2–3 weeks since it may derive from processes started in the open bay even before December. Further elements supporting this interpretation are obtained from the multivariate statistical analysis of data of the 1987–88 and 1989–90 seasons and reported below.

Differently from other oceans, the polar seas offer a unique opportunity to observe vertical metal distribution when the photosynthetic activity is negligible due to the absence of light (or low illumination) and the presence of marine ice. The present results in Antarctica show that at the beginning of the summer, and presumably in the winter, the cadmium vertical profile is completely homogeneous. This observation gives further evidence in favour of the general interpretation of Cd surface depletion in the Oceans in terms of its involvement in the biological cycles[17].

It is interesting to note that during the period of observation, the Cd concentration in the surface water is decreased by about one order of magnitude; in particular it passed from about 0.8 nmol/l in the first period of observation (November–December) to 0.19, 0.16 and 0.10 nmol/l on January 6, January 30 and February 11, respectively (see Figure 5, 1990–91 campaign). This high seasonal variability can explain the different values obtained in different years,

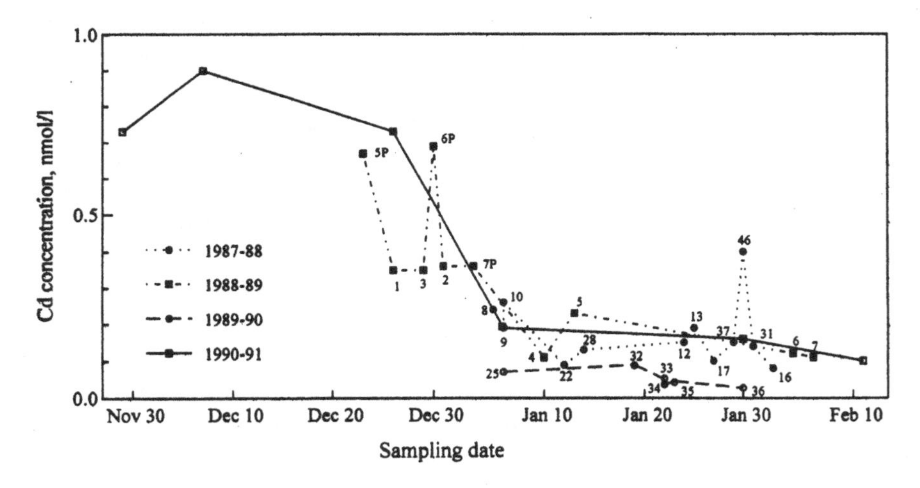

FIGURE 5 Summer variation of cadmium surface concentration in Terra Nova Bay during four successive campaigns.

when samples were collected in different periods of the seasons or when the seasonal evolution of biological activity was brought forward or delayed due to different meteorological and climatic conditions which may occur from year to year.

To clarify this aspect further the results obtained in previous expeditions are added in Figure 5. From this plot it can be observed that the surface cadmium concentration in Terra Nova Bay follows approximately the same decreasing temporal trend in each season irrespective of the spatial location of samples in the bay and of the year of observation. The only deviation seems to be represented by the data obtained in the summer of 1989–90, when the observed concentrations were about half the mean values obtained on the same date during the two previous seasons. Actually the 1989–90 season was exceptional from a climatic point of view, with an early temperature increase[39] (by about 15 days) and a very intensive phytoplankton bloom[40] (see also the subsequent discussion on the 1987–88 and 1989–90 seasons). As a consequence the points in Figure 5 referring to the 1989–90 season should probably be considered as needing to be shifted to the right with respect to the actual sampling date. It can therefore be concluded that, as a general rule, the temporal factor appears dominant for determining the Cd concentration level and for explaining its variation observed either in the same season or from year to year.

Lead

Data reported in Table II show that the lead concentrations in the water column vary from 0.024 to 0.132 nmol/l. These values are comparable with literature data reported for the Weddell Sea[11] (0.015–0.062 nmol/l) and for the Weddell and Scotia Seas[8] (0.010–0.103 nmol/l); moreover they are consistent with upwelling antarctic areas subject to the influence of volcanic and hydrothermal activities (Mt. Melbourne and Mt. Erebus volcanoes are located in the area) and with the presence of polynhyas[8].

The variations recorded during the campaign (Figure 6) are generally lower than those observed for cadmium. The effect of both the high analytical variability, due to the extremely low concentration involved, and the very occasional contamination of the samples also superimposes on the actual lead variation. Consequently the seasonal behaviour of lead is registered in a less marked manner. Nevertheless a few interesting considerations can be made on the basis of the data.

The surface concentration, practically constant to about 0.1 nmol/l from November 29 to December 26, then decreases rapidly to 0.065 nmol/l on January 6;

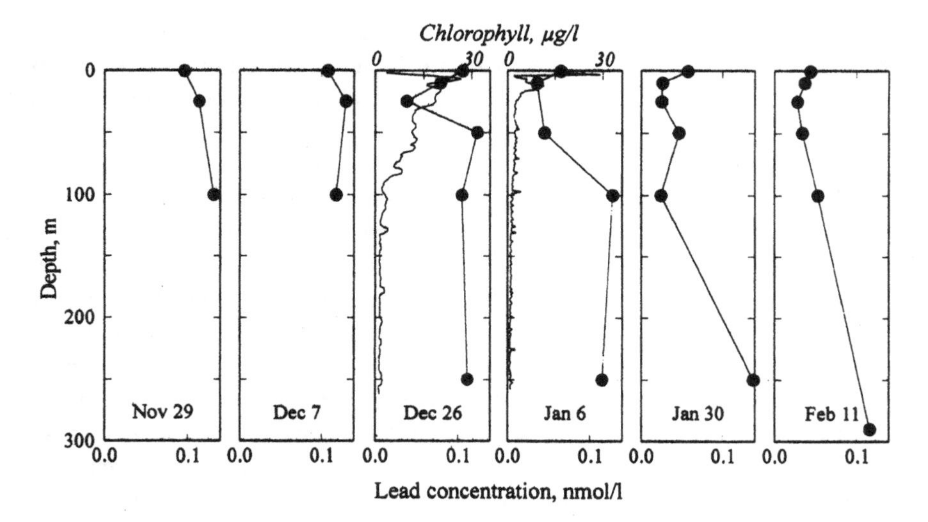

FIGURE 6 Seasonal evolution of depth profile for lead concentration (•) and chlorophyll (—). 1990–91 campaign.

subsequently a more gradual decline is observed to reach the value of 0.044 nmol/l on February 11 (see Figure 7). This interval includes practically all the values obtained in the same area (Gerlache Inlet) during previous expeditions, when samples were collected at different moments of the climatic and biological evolution of the summer season (in nmol/l 1987–88: 0.025–0.114[1]; 1988–89:

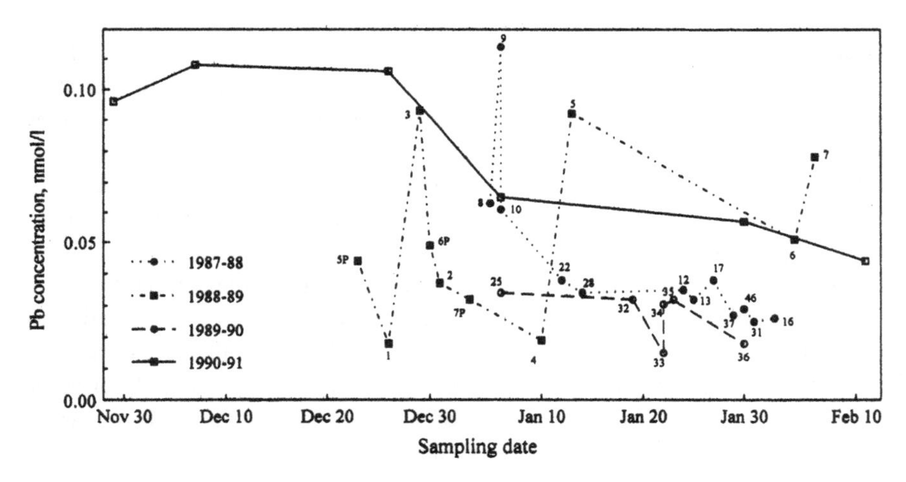

FIGURE 7 Summer variation of lead surface concentration in Terra Nova Bay during four successive campaigns.

0.018–0.093[6]; 1989–90: 0.015–0.044[6]). In the case of lead it is to be remembered that different concentration levels have already been recognised (1987–88 season) between the inner area of Gerlache Inlet (0.061–0.114 nmol/l) and the open sea of Terra Nova Bay (0.025–0.038 nmol/l)[1]. Thus the spatial effect cannot be neglected in comparing data obtained in different years from different locations. In particular this effect is responsible for the discrepancies in the lowest values observed between the results of the present work (1990–91 season) and those previously reported (see above). Nevertheless the plot of all the data against the date of collection (Figure 7) also confirms a general decreasing trend during the summer for lead, although to much smaller degree than cadmium.

A rapid depletion of lead in Antarctic surface seawater has also been emphasized for the Weddell and Scotia Seas and interpreted in terms of particulate scavenging processes during periods of intense primary production in the austral summer[8].

As regards the deep water (Figure 6) it can be observed that the vertical profile of lead concentration appears practically uniform at the beginning of the season, with values at the same level as those of the surface water, while a significant depletion of the concentration begins to occur in the subsurfacial layers when the biological activity is increasing rapidly (see chlorophyll profile in Figure 6). At the end of the process the Pb concentration in these layers is reduced to about one third of the initial value and, differently from what happen for Cd, the depletion extends deeper and deeper during the season, down to at least 100 m.

It has been observed that a surface enrichment is typical of the oceans subject to atmospheric input originating from remote anthropogenic sources or of coastal areas subject to local immission[17]. From the present study the absence of a marked surface increase of metal concentration with respect to the deep baseline values indicates the absence in Terra Nova Bay of significant Pb pollution coming from remote or local areas.

Comparison of the 1987–88 and 1989–90 Seasons

The combined effects of physical and biological processes in determining metal distributions in the surface waters of Terra Nova Bay together with evidence of year to year variations with respect to the summer evolution of biological activity can be exploited better by comparing data collected during the 1987–88 and 1989–90 expeditions. A multivariate statistical approach is used for this comparison. Unfortunately a general comparison with the other seasons investigated, including 1990–91, and the analysis for deep waters are not possible due to the incompleteness of data.

The variables used for the statistical analysis are the concentrations of cadmium and lead[1,2,5,6], copper[4], silicates, phosphates, nitrates and nitrites[22,23], chlorophyll[24–26], temperature[24,27,28] and salinity[23,25,28]. Principal Component Analysis (PCA)[29] is used mainly to elucidate the variable association and the group structure in the data set. The first three, varimax rotated, principal components (cumulative explained variance 82.0%) are considered relevant for the study and interpretation of the data.

PCA results, presented in terms of biplots[29,41] in Figure 8, show clear associations between nutrients (except nitrites), salinity and cadmium on the first PC, and metals (including Cd) on the third PC, while the second contrasts temperature and chlorophyll.

Considering that nutrients in surface waters (as described above) are consumed by productive activity, and that variation of salinity is due to physical processes, it follows that the first PC accounts for both processes, which cannot be separated and which together contribute to the variability of cadmium concentration (at least in part). These results are in agreement with the interpretation of cadmium data given above.

The scatter plot of data on principal components (Figure 8) shows a net separation of the 1989–90 samples from those of 1987–88 confirming the overall different behaviour of the two seasons as pointed out by biological studies (see above)[40]. The differentiation is observed mainly along the first PC then determined for the most part by nutrient content, salinity and cadmium. The only exception is sample 25, which was collected at a depth of 20 m (all the others are surface water). A subgroup consisting of samples 8, 9, 10 and 46 can be observed in the 1987–88 data set. This differentiation, which emerges when the third PC is considered (explained variance 12.2%) and is mainly determined by the Pb, Cu and Cd content, has already been recognised previously in terms of spatial location of samples[1,4].

DISCUSSION

Metal concentration in oceanic waters is governed by the balance between their rate of addition and their rate of removal. The major sources of trace metals in sea water derive from (i) atmospheric or riverine input and (ii) the interaction of waters with newly formed oceanic crustal material and hydrothermal activity[42]. Trace elements mobilised from human activities and transported via atmosphere

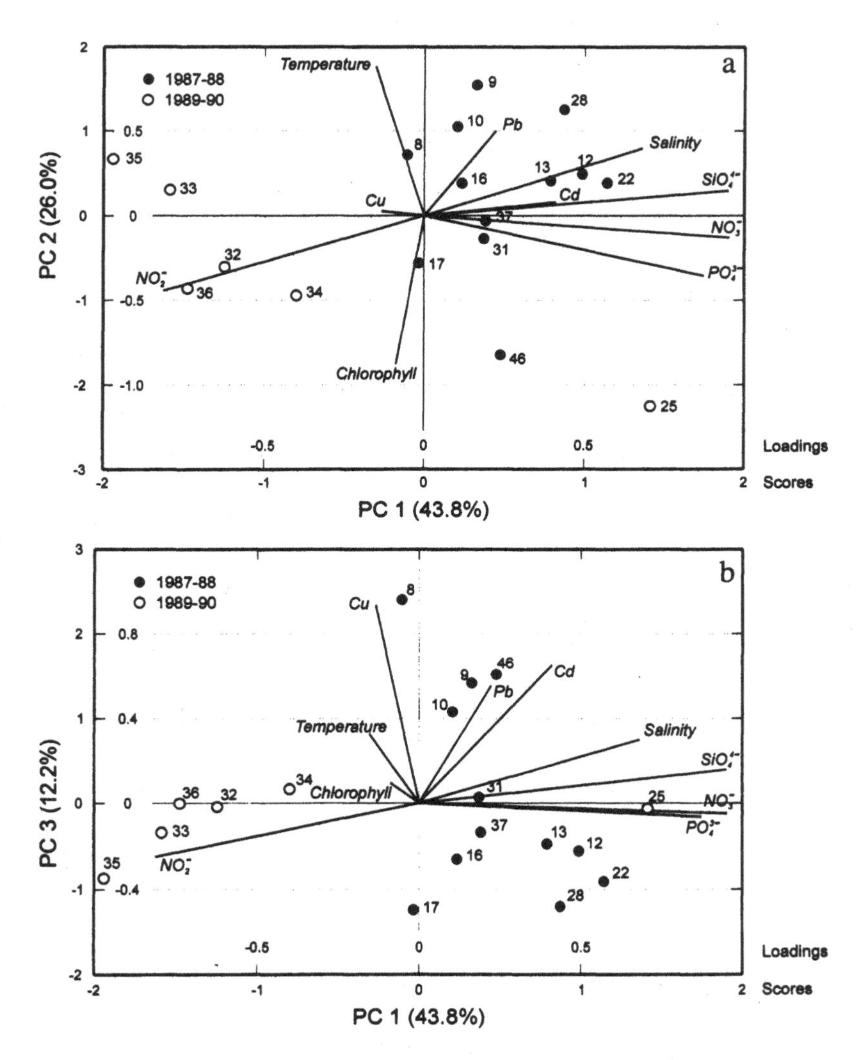

FIGURE 8 Principal component biplot (after varimax rotation) with respect to (a) the first two PCs and (b) the 1st and 3rd PCs. Data from 1987–88 and 1989–90 expeditions.

can contribute significantly to their distribution in the oceanic regions[17]. For most elements the main removal process is via sedimentary deposition and burial; however, before being ultimately removed from the sea water reservoir to the sediments, they can undergo varying degrees of internal recycling as a result of biogeochemical processes[43]. The recycling can involve the uptake of trace elements by particulate phases produced principally by phytoplankton in the photic layer and regeneration into the deep ocean where the particles are largely redissolved as a consequence of respiration and oxidation processes. Biological cycles appear to be responsible for the transport of a wide range of elements from the surface to the deeper layers[17,42,43].

The processes of particle production and dissolution are well-separated in space and time, and much can be understood about particle-water interactions from the observation of vertical profiles of the dissolved and particulate constituents in coastal and oceanic waters[42]. In particular studies of the seasonal evolution of vertical profiles of the elements and their relationship with other oceanographic parameters (e.g., temperature, salinity, nutrients) carried out in areas with uncomplicated hydrodynamic processes (for example in oligotrophic subtropical gyres characterised by minimal vertical mixing and low productivity), have enabled processes to be elucidated which control the element distribution[13,17,42,44,45].

A rough measurement of the intensity of the particle-water interactions for different elements can be the mean oceanic residence times[42]. On the basis of these and the consequent patterns of the vertical profiles, the elements can be subdivided into accumulated, recycled and scavenged elements[17]. Accumulated elements (residence time $>10^5$ years) exhibit a higher concentration in sea water relative to their crustal abundance and a uniform vertical profile (when normalised to sea water salinity). Recycled elements (intermediate residence time, between 10^3 and 10^5 y) tend to be involved in the biological particle cycle; their vertical profile presents a significant surface depletion and deep regeneration (typical nutrient profile). Scavenged elements (residence time $<10^3$ y) interact strongly with particulate matter and are removed on time scales shorter than a single stirring cycle of the ocean reservoir. These elements, present in sea water at very low concentrations relative to their crustal abundance (especially in deep water), exhibit different profiles according to their primary input; normally a surface maximum is displayed because they enter the sea via aeolian dust (possibly enriched due to anthropic influence) or from the ocean margin.

Interpretation of data obtained in the present work should account for the different geochemical characteristics of cadmium and lead, the different interaction processes of metals with phytoplankton and particles and the seasonal evolution of biological processes during the investigation.

Cadmium and lead are typical recycled and scavenged metals, respectively[17]. At the very beginning of the summer season, when both biological activity and new formation particle density are at a minimum, both metals exhibit homogeneous vertical distributions at the higher concentration levels; these are consistent with a coastal region with an upwelling hydrodynamic regime, reduced processes of biological uptake and scavenging and without high local sources. The presence of a polynya in the area of Terra Nova Bay[33,35] highlights vertical flows in the winter and the consequent homogeneization of the water column. Concentration values are comparable with those observed in the coastal Pacific Ocean[17]. Subsequently a general decrease in metal content is observed both in the surface and in the shallow waters. The amount of this depletion, the temporal trend and the layer which is the most affected by the phenomenon (as shown by the shape of the deep profile and its change during the season) are remarkably dependent on the metal considered.

Cadmium, as a nutrient-type element, interacts with particulate matter, being subject to uptake from biological systems and subsequently discarded with detritus and fecal pellets. A quick depletion of the metal is observed in the euphotic layer when the phytoplankton bloom starts. This depletion increases during the summer season but it remains confined to approximately the same layer and does not affect deeper strata. Note that the highest chlorophyll concentrations in the Terra Nova Bay are located in the 20–30 m upper layer (see results in Figure 4 and in Innamorat et al.[46]).

Lead, a scavenged metal, interacts with particulate by adsorption and reaches the sediment in association with the host particles of the highest dimensions. These particles are generated from chemical reactions, biologically mediated, tending to destroy their organic components and compacting the smaller particles into a larger one. Consequently the sedimentation process for this metal is maximum when the number of particles of higher dimensions is maximum[41].

The suspended matter in the Ross Sea originates mainly from biological activity[46,47] and the maximum density of the largest particles is located deeper with respect to the zone of maximum production. In the coastal area the largest particles present a more homogeneous vertical distribution than the smaller ones and they reach the maximum concentration during a late or senescent phase of a bloom[47]. This explains the slow decreasing trend of Pb concentration in the vertical profile starting in December which became more consistent and extended to almost all the vertical profile in February, when particles of higher dimension became prevalent.

Finally it is to be noted that the temporal trend of the Pb profile observed here is in agreement with observations carried out in the Pacific, where the residence time of lead was evaluated as about two years in surface waters and several hundred years in the deep ocean[45].

CONCLUSIONS

Measurements carried out in Antarctica in different periods of the summer and at different depths have confirmed that DPASV is sufficiently sensitive to be used for direct determination of Cd and Pb in Antarctic seawater, thus avoiding lengthy enrichment procedures which are susceptible to sample contamination. This feature, together with the easy transport-capability of the instrumentation (low cost, small size, low weight, minimal facilities required) allows determinations to be carried out in the remotest areas of the world, like research stations in Antarctica or ice-breaking ships in polar regions, provided that a clean chemistry laboratory is available. The ability to obtain reliable measurements on site immediately after seawater collection is of particular relevance to checking samples and sampling procedures (modified or adjusted in case of contamination) and to providing reference values to be compared with later results obtained in the laboratory after a long storage period.

The seasonal evolution obtained for the metal content along the water column of Terra Nova Bay during the 1990–91 austral summer emphasizes the substantial difference in the mechanisms which control the distribution of Cd and Pb in seawater. The prevalent interactions of cadmium with biological systems and of lead with particulate matter are also hypothesized in the polar coastal area studied. A definitive clarification of this aspect requires an ad hoc campaign planned to obtain more frequent observations and contextual, detailed information from both the hydrological and hydrodynamical points of view. The seasonal variation observed at the surface explains the variability of metal concentration observed in the previous three expeditions in the same area.

The vertical distributions of cadmium concentration are homogeneous before both the pack defrosting and the phytoplankton bloom start in the whole area; afterwards, in a few weeks, biological activity generates the typical nutrient profile characteristic of cadmium, with a surface variation which confirms the trend observed in a previous study[2]. An apparently similar behaviour is observed for lead; in this case, however, a lower and delayed effect is noted in the concentration depletion which nevertheless extends to deeper layers during the season than that for cadmium.

Again with respect to lead it is important to observe that the vertical profile does not show the high surface increase typical of anthropized areas[44], so serious problems of pollution deriving from local or remote sources can be excluded. This conclusion is in agreement with lead data obtained for snow in Victoria Land which showed a drastic reduction of the lead content during the last decade to a value typical of the beginning of the century[48].

At present other aliquots of the same samples considered here (transported frozen to Italy) are being subjected to laboratory analysis for metal determination and speciation according to the methodology already used in the same area[1,2,4]. As well as the obvious aim of confirmation of the general findings from stored samples, the new data will enable discussion of the aspects of metal distribution during the summer in terms of metal speciation. Finally a further contribution to the same subject is expected from the results of a similar investigation now in progress in Wood Bay, an area more distant from the Italian Station than Terra Nova Bay.

Acknowledgements

This work was supported by the Italian National Research Programme in Antarctica (PNRA). Thanks are due to the Environmental Impact-Chemical Methodologies researchers for their accurate work in the field during the 1990–91 expedition, to all the members of the Italian technical staff for their assistance at sea and to C. Zago and A. Gambaro for the helpful technical assistance in the laboratory.

The authors are grateful to colleagues in the "Physical, Chemical and Biological Oceanography" Sector for providing the oceanographic data collected during the 1987–88 and 1989–90 campaigns, and to A. Bergamasco for the useful discussions on the hydrography of the Ross Sea.

An anonymous referee is gratefully acknowledged for the stimulating comments which enabled us to improve the manuscript significantly.

References

[1] G. Capodalglio, G. Toscano, G. Scarponi and P. Cescon, *Ann. Chim.*, **79**, 543–559. (1989). Correction, *Ann. Chim.*, **80**, 393 (1990).
[2] G. Capodaglio, G. Scarponi, G. Toscano and P. Cescon, *Ann. Chim.*, **81**, 279–296 (1991).
[3] G. Capodaglio, G. Scarponi and P. Cescon, *Anal. Proc.*, **28**, 76–77 (1991).
[4] G. Capodaglio, G. Toscano, G. Scarponi and P. Cescon, *Intern. J. Environ. Anal. Chem.*, **55**, 129–148 (1994).
[5] G. Capodaglio, G. Toscano, G. Scarponi, C. Barbante, C. Turetta and P. Cescon, *Methodological aspects of cadmium speciation in seawater by Anodic Stripping Voltammetry*, in preparation.
[6] G. Scarponi, G. Capodaglio, C. Barbante, C. Turetta and P. Cescon, *Lead speciation in Antarctic seawater*, in preparation.
[7] R. F. Nolting and H. J. W. de Baar, *Mar. Chem.*, **45**, 225–242 (1994).
[8] A. R. Flegal, H. Maring and S. Niemeyer, *Nature*, **365**, 242–244 (1993).
[9] S. Westerlund, and P. P. Öhman, *Geochim. Cosmochim. Acta*, **55**, 2127–2146 (1991)
[10] P. M. Saager, H. J. W. de Baar and R. J. Howland, *Deep-Sea Res.*, **39**, 9–35 (1992).
[11] L. Mart, H. Rutzel, P. Klahre, L. Sipos, U. Platzek, P. Valenta and Nurnberg, H. W. *Sci. Total Environ.*, **26**, 1–17 (1982).
[12] M. J. Orren and P. M. S. Monteiro, In: *Antarctic Nutrient Cycles and Food Webs* (W. R. Siegfried, P. R. Condy and R. M. Laws, eds., Springer-Verlag, Berlin), pp. 30–37 (1985).
[13] K. W. Bruland, *Earth Planet. Sci. Lett.*, **47**, 176–198 (1980).
[14] K. W. Bruland and R. P. Franks, In: *Trace Metals in Seawater* (C. S. Wong, E. Boyle, K. W. Bruland, J. D. Burton and E. D. Goldberg, eds., Plenum Press, New York), pp. 395–414 (1983).
[15] K. W. Bruland, K. J. Orians and J. P. Cowen, *Geochim. Cosmochim. Acta*, **58**, 3171–3182 (1994).

[16] R. D. Frew and K. A. Hunter, *Nature*, **360**, 144–146 (1992).
[17] K. W. Bruland, In: *Chemical Oceanography* (J. P. Riley and R. Chester, eds., Academic Press, London), Vol. 8, Chapt. 45, pp. 157–220 (1983).
[18] G. Scarponi, G. Capodaglio, C. Barbante, and P. Cescon, In: *Element Speciation in Bioinorganic Chemistry* (S. Caroli, ed., Wiley, New York), Chapt. 11, pp. 363–418 (1996).
[19] G. Capodaglio, G. Toscano, P. Cescon, G. Scarponi and H. Muntau, *Ann. Chim.*, **84**, 329–345 (1994).
[20] G. Capodaglio, G. Scarponi, G. Toscano, C. Barbante and P. Cescon, *Fresenius J. Anal. Chem.*, **351**, 386–392 (1995).
[21] L. Mart, H. W. Nurnberg and P. Valenta, *Fresenius Z. Anal. Chem.*, **300**, 350–362 (1980).
[22] G. Catalano and F. Benedetti, In: *Oceanographic Campaign 1987–88. Data Report. Part I.* (National Scientific Commission for Antarctica, Genova), pp. 61–83 (1990).
[23] G. Catalano, F. Benedetti and M. Iorio, In: *Oceanographic Campaign 1989–90. Data Report. Part I.* (National Scientific Commission for Antarctica, Genova), pp. 25–32 (1991).
[24] M. Fabiano, P. Povero, G. Catalano and F. Benedetti, In: *Oceanographic Campaign 1989–90. Data Report. Part I.* (National Scientific Commission for Antarctica, Genova), pp. 35–71 (1991).
[25] M. Innamorati, G. Mori, L. Lazzara, C. Nuccio, M. Lici and S. Vanucci, In: *Oceanographic Campaign 1987–88. Data Report. Part I.* (National Scientific Commission for Antarctica, Genova), pp. 161–238 (1990).
[26] M. Innamorati, L. Lazzara, G. Mori, C. Nuccio and V. Saggiomo, In: *Oceanographic Campaign 1989–90. Data Report. Part I* (National Scientific Commission for Antarctica, Genova), pp. 141–252 (1991).
[27] A. Boldrin and C. Stocchino, In: *Oceanographic Campaign 1987–88. Data Report. Part I.* (National Scientific Commission for Antarctica, Genova), pp. 11–57 (1990).
[28] A. Artegiani, R. Azzolini, E. Paschini and S. Creazzo, In: *Oceanographic Campaign 1989–90. Data Report. Part II.* (National Scientific Commission for Antarctica, Genova), pp. 5–62 (1992).
[29] I. T. Jolliffe, *Principal Component Analysis* (Springer-Verlag, New York), 271 pp. (1986).
[30] UNISTAT Statistical Package, Version 3.0, Unistat Ltd., London, 1994.
[31] G. L. Pickard and W. J. Emery Descriptive Physical Oceanography. An Introduction. Pergamon Press, Oxford (1990).
[32] S. S. Jacobs, A. F. Amos and P. M. Bruchausen, *Deep-Sea Res.*, **17**, 935–962 (1970).
[33] S. S. Jacobs, R. G. Fairbanks and Y. Horibe, In: "Oceanology of the Antarctic Continental Shelf", Jacobs S. S. (Ed.), A. G. U., *Antarctic Research Series*, **43**, 59–85 (1985).
[34] V. V. Klepikov and Yu. A. Grigoriev *Inf. Byull. Sov. Antarkt. Eksped.*, **56**, Engl. Trasl., *Information Bulletin at the Soviet Antarctic Expedition*, **6**, 52–54 (1996).
[35] C. Stocchino and G. M. R. Manzella, Report National Research Council of Italy CNR, Area della Ricerca di Genova, 56 pp. (1991).
[36] C. Stocchino and C. Lusetti, *Ist. Idrogr. della Marina, Genova*, F.C. 1132, 55 pp. (1990).
[37] S. S. Jacobs, A. L. Gordon and A. F. Amos, *Nature*, **277**, 469–471 (1979).
[38] A. C. Redfield, B. H. Ketchum and F. A. Richards, In: *The sea* (N. M. Hill, ed., Wiley, New York), pp. 26–77 (1963).
[39] Italian National Research Programme for Antarctica (PNRA), *Report of Automatic Weater Stations* (ENEA, Rome, Casaccia PO Box 2400, Years from 1987 to 1991).
[40] M. Innamorati, L. Lazzara, G. Mori, C. Nuccio and V. Saggiomo, In: *Proc. 9° Congr. Assoc. Ital. Oceanol. Limnol.* (G. Albertelli, W. Ambrosetti, M. Piccazzo and T. Ruffoni Riva, eds., A.I.O.L., Genova), pp. 605–612 (1992).
[41] K. R. Gabriel, *Biometrika*, **58**, 453–467 (1971).
[42] R. Chester, *Marine Geochemistry* (Unwin Hyman, London), 698 pp (1990).
[43] M. Whitfield and D. R. Turner, In: *Aquatic Surface Chemistry*; *Chemical Processes at the Particle-Water Interface* (W. Stumm, ed., Wiley, New York), Chapt. 17, pp. 457–493 (1987).
[44] B. K. Schaule and C. C. Patterson, In: *Trace Metals in Seawater* (C. S. Wong, E. Boyle, K. W. Bruland, J. D. Burton and E. D. Goldberg, eds., Plenum Press, New York), pp. 487–504 (1983).
[45] E. A. Boyle, S. S. Huested and S. P. Jones, *J. Geophys. Res.*, **86**, 8048–8066 (1981).
[46] M. Innamorati, L. Lazzara, L. Massi, G. Mori, C. Nuccio and E. V. Saggiomo, In:

Oceanografia in Antartide (V. A. Gallardo, O. Ferretti and H. I. Moyano, eds., Proc. Intern. Symp., Conception, Chile), pp. 235–252 (1991).

[47] L. Lazzara and C. Nuccio, In: *Proc. 10° Congr. Assoc. Ital. Oceanol. Limnol.* (G. Albertelli, R. Cattaneo-Vietti and M. Piccazzo, eds., A.I.O.L., Genova), pp. 655–680 (1994).

[48] C. Barbante, C. Turetta, G. Capodaglio and G. Scarponi, *Intern. J. Environ. Anal. Chem.*, submitted.

RECENT DECREASE IN THE LEAD CONCENTRATION OF ANTARCTIC SNOW

C. BARBANTE[a,b], C. TURETTA[b], G. CAPODAGLIO[a,b] and G. SCARPONI[c*]

[a]Department of Environmental Sciences, University of Venice Ca' Foscari, S. Marta 2137, I-30123 Venice, Italy; [b]Study Centre for Environmental Chemistry and Technology-CNR, University of Venice Ca' Foscari, S. Marta 2137, I-30123 Venice, Italy; [c]Faculty of Mathematical, Physical and Natural Sciences, University of Ancona, Via Brecce Bianche, I-60131 Ancona, Italy

Differential Pulse Anodic Stripping Voltammetry (DPASV) was applied to determine the lead concentration in recent snow at two sites in the Victoria Land region, East Antarctica. Snow samples were collected during the 6th Italian Scientific Expedition to Antarctica (austral Summer 1990–91) along the wall of 2.5 m-deep hand-dug pits and by coring to a depth of about 11 m. The measurements revealed that lead content in Antarctic snow increased continuously from 1965 (about 3 pg/g) to the early 1980s (maximum about 8 pg/g), after which a marked, rapid decrease took place during the second half of 1980s, down to 2–4 pg/g in 1991. Estimates of the lead contributions from rocks and soils, volcanoes and the marine environment, together with analysis of statistical data on non-ferrous metal production and gasoline consumption, and the corresponding lead emissions into the atmosphere of the Southern Hemisphere, show that a net anthropogenic component is present and support the hypothesis that the trend observed in Antarctic snow may be related to lead consumption in gasoline, which firstly was on the rise, then declined owing to the increased use of unleaded gasoline.

Keywords: Lead; DPASV; snow; Victoria Land; Antarctica

INTRODUCTION

The massive use of lead in human activities for the manufacture of a variety of products, such as batteries, pigments, rolled and extruded items, cable sheaths and above all gasoline additives in the form of highly volatile alkyl lead compounds, causes large emissions of the metal into the environment and poses worrying ecological and health problems which demand an assessment of the consequent changes in the large-scale atmospheric cycles of this metal[1].

*Corresponding author. Fax No.: +39-41-2578549.

The analytical investigation of polar ice caps (the earth's ice archives) has proved to be one of the most powerful ways of obtaining information on present, recent and past changes in the earth's atmosphere, including heavy metal changes[2,3].

The lead concentration vertical profile of the Greenland ice cap revealed that this metal is one of the earliest polluting substances of the Northern Hemisphere as used by the Greek and Roman civilizations, long before the Industrial Revolution[4]. In more recent periods the same profile clearly demonstrated the increase in large-scale pollution of the atmosphere from the Industrial Revolution, particularly after the Second World War and up to the late 1960s and the recent decrease mainly associated with the reduction in the use of lead in gasoline[5–9].

Recent studies have proved that even the remote Antarctic continent is sig nificantly contaminated by lead, possibly due to the same anthropic sources[10–15], while a few data sets account for a possible, more recent, declining trend[14–16].

As a further contribution to the knowledge of recent changes of lead concentration in Antarctic snow, we present here new data obtained from samples collected at two sites in Victoria Land during the VI Italian Expedition to Antarctica (Summer 1990–91). Direct determination of lead in melted, decontaminated samples was carried out by Differential Pulse Anodic Stripping Voltammetry (DPASV), thus avoiding any long and contamination-prone enrichment procedure.

EXPERIMENTAL

Laboratories and Chemicals

Both warm and cold (−20°C) clean chemistry laboratories were available with Class 100 vertical laminar flow areas. Wood benches with polypropylene shelves or Teflon coated stainless steel tables were installed in the laboratories. The cleaning of materials, handling of samples and instrumental measurements were carried out in the warm laboratory while the treatment and decontamination of frozen samples were performed in the cold laboratory. Researchers followed the clean room procedures strictly in both these clean environments.

Ultrapure water was obtained from Milli-Q or Alpha-Q systems (Millipore, MA, USA). NIST (National Institute of Standards and Technology, Gaithersburg, MD, USA) ultrapure HCl (32%) and HNO_3 (70%)[17] were used, respectively, for acidification of samples and for the final steps of the cleaning procedures of plastic items (see below). Analytical grade HNO_3 65% (Merck, Darmstadt, Germany) was used during the first step in the cleaning of materials. Saturated KCl solution (Suprapur KCl, Merck, ultrapure water) was further purified by passing it through Chelex 100 chelating resin (Bio-Rad, CA, USA). The 2.5 × 10^{-2} M $Hg(NO_3)_2$ solution was prepared by oxidation of hexadistilled Hg with concentrated ultrapure HNO_3 and diluting with ultrapure water. Nitrogen of chromatographic grade (purity >99.999%) was obtained through a nitrogen generator (Model NG 4000, Claind, Italy) or in high pressure cylinders (SIAD, Italy). Detergent powder from Alconox (NY, USA) was used.

Standard solutions (Pb: 10 μg/l; Al: 1, 5, 10 μg/l) were prepared through successive dilutions of AAS stock solutions (Titrisol, 1000 ppm, Merck). NASS-3 and NASS-4 open ocean seawater reference materials for trace metals were obtained from the National Research Council of Canada[18] and were used to check the analytical procedure for accuracy.

Cleaning of Materials

The PFA sample containers, the electrochemical cell components, Teflon and polyethylene bottles, and scrubbers, hammers and chisels used to decontaminate the pit walls and snow cores (see below) were acid cleaned following a five-step procedure previously described in detail[15].

Briefly items were cleaned as follows. Conventional chemical laboratory: immersion in detergent bath (40 °C, 10 days), rinsing with ultrapure water, immersion in 1:10 diluted HNO_3 analytical grade (40 °C, one week), repeated careful rinses with ultrapure water. Clean chemistry laboratory: immersion in 1:100 diluted ultrapure HNO_3 (35 °C, 15 days) rinsing and immersion in 1:1000 diluted ultrapure HNO_3, (35 °C, 20 days); finally, bottles are rinsed, filled with a 1:1000 diluted ultrapure HNO_3 fresh solution and stored inside double polyethylene bags, while scrubbers, knives and other items remain in the last bath until use.

Sampling and Datation

In January 1991, during the VI Italian Scientific Expedition to Antarctica, snow samples were collected at two sites in Victoria Land down to a depth of about 11 m (see Figure 1). The first site was located 40 km North-West of the Terra Nova Bay Station near the McCarthy Ridge (74° 32′ S – 162° 56′ E, elevation 700 m, 40 km from the sea, mean snow accumulation rate 27 g cm^{-2} y^{-1}[19]); the second was located 100 km North of the Station on the Styx Glacier plateau (73° 52′ S – 163° 42′ E, elevation 1700 m, 50 km from the sea, mean snow accumulation rate 16 g cm^{-2} y^{-1}[19]).

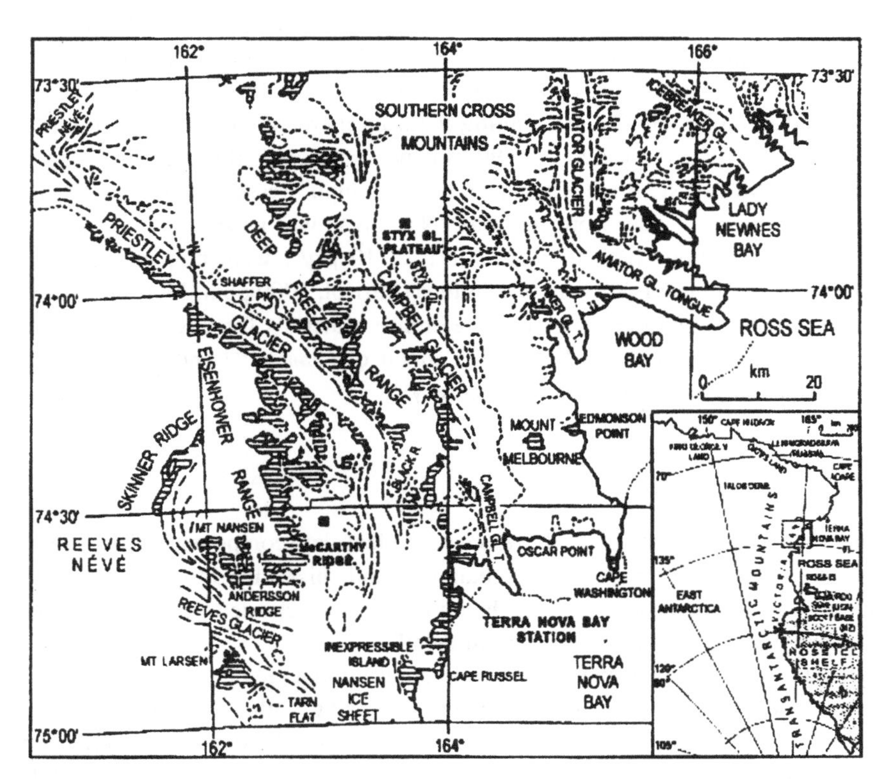

FIGURE 1 Locations of sampling sites on McCarthy Ridge and the Styx Glacier plateau (Victoria Land, East Antarctica). Ice-free area hatched.

Shallow samples were collected in 2.5-m deep hand-dug pits, inserting cylindrical containers horizontally in the carefully decontaminated upwind pit wall. After collection the containers were sealed inside double polyethylene bags and stored frozen (−20°C) until analysis. Pre-cleaned Teflon PFA containers with i.d. 11 cm, length 26 cm, capacity 2 litres (Savillex, MN, USA) were used. During the sampling operations researchers wore clean room garments, masks, polyethylene gloves and boot covers to prevent any contamination problem.

Deep samples were collected by means of a fibre-glass hand auger (i.d. 10.4 cm, length of the core barrel 127 cm, PICO, Polar Ice Coring Office, University of Nebraska-Lincoln, NB, USA). Snow core sections from 15 to 78 cm were obtained; they were sealed in two polyethylene bags and stored frozen (−20 °C) in rigid PVC tubes.

The age of the pit samples was estimated from the hydrogen peroxide deep profiles obtained in the same pits and already reported[19]. As usual[20] the H_2O_2

summer concentration maxima were used to reconstruct the age of snow layers. Each pit sample, considering the container diameter and the different thickness of the annual layers, spanned between 2 and 4 months. It is to be noted that the use of cylindrical samplers favours the snow layers in the central part of the cross section with respect to the upper and lower parts.

The age of the snow cores was estimated from the hydrogen peroxide and $\delta^{18}O$ vertical profiles measured in cores drilled at the same sites in the 1991–92 season[21] and considering the known snow accumulation during 1991 (41.0 cm and 29.5 cm of snow for McCarthy Ridge and Styx Glacier plateau, respectively)[22]. The uncertainty in datation was about 0.5 years. The analysed part of the snow core sections (usually 20 cm in length, see below) covers a period of about 4 months for McCarthy Ridge and from 4 to 10 months for the Styx Glacier plateau.

The depths at which analysed samples were collected and the corresponding estimated ages are reported below in Table I.

Decontamination of Snow Cores

Despite the great care taken in the field to avoid contamination, the external layers of the snow cores are more or less contaminated by metal impurities due to the auger and the long storage period. For this reason a special decontamination procedure was carried out which enabled concentric layers of snow to be separated and lead concentration to be measured from the outside to centre in order to obtain radial concentration profiles.

Five snow core sections (McCarthy Ridge: 308–328 and 328–348 cm; Styx Glacier plateau: 668–688, 990–1010 and 1010–1030 cm) were decontaminated in the Laboratoire de Glaciologie et Geophysique de l'Environnement (LGGE), Grenoble, inside a laminar flow bench installed in a cold laboratory (−20°C) according to a procedure already described in detail[12]. Briefly, the snow core extremities are removed to prepare a 20-cm section with untouched surfaces; this section is inserted inside a 10.4 cm diameter polyethylene beaker. Then an ultra-clean polyethylene beaker (5.0 cm i.d.) is hammered longitudinally in the central part of the core and a series of 12 smaller polyethylene beakers (i.d. 1.0 cm) are hammered close to it. Finally the remaining external material is also collected. The procedure allows samples from three concentric layers to be obtained for each section, corresponding to the following ranges in radius: central core 0–2.5 cm, intermediate layer 2.5–3.5 cm, external layer 3.5–5.2 cm.

The remaining core sections were decontaminated in our cold, clean laboratory following the chiseling procedure already described in the literature[23]. Briefly, the snow core section is fixed on the Teflon holders of a clean polyeth-

Table I. Lead concentrations and estimated deposition period for samples collected on the McCarthy Ridge and Styx Glacier plateau.

Depth[a] (cm)	Deposition period[b]	Pb concentration (pg/g) Measured values	Mean
		McCarthy Ridge	
5	S 1990-91	4.3, 5.1, 3.6	4.3 [c]
200	W 1987	7.0, 7.0, 6.5	6.8 [c]
308-328	1985	[d]*(213[e]) (11.8) (25.4)*	11.8 [f]
328-348	1984	*(96, 100, 89)* (7.9) (4.6, 6.5)	6.8
		Styx Glacier plateau	
5	S 1990-91	53	53 [f]
15	S 1990-91	2.1, 2.4, 2.3	2.3 [c]
25	W 1990	2.6, 2.1, 2.2	2.3
48	S 1989-90	3.3, 2.6, 3.4, 3.0	3.1
70	W 1989	3.3, 4.6, 3.2	3.7
100	S 1988-89	7.0, 5.3, 7.8	6.7
150	S 1986-87	8.5, 7.2, 6.0	7.2
200	S 1985-86	13.8, 15.3, 9.7, 12.4	12.8 [c, f]
133-153	1989	*(18)* (-[g]) (-) (4.5) (4.5, 3.8)	4.4
219-239	1987	*(112)* (-) *(17.8)* (9.2) (6.8, 6.7)	8.0
267-287	1986	*(170)* (-) *(31)* (9.1) (5.6, 5.8)	7.4
296-316	1984	*(374)* (-) *(33.8) (18.8)* (6.7, 5.0)	5.8
325-345	1983	*(389)* (-) *(27) (7.8)* (5.5, 6.1)	6.8
350-370	1982	*(164)* (-) *(12.1)* (8.8) (6.2, 8.5)	8.1
438-458	1981 [h]	[h]*(202)* (-) *(13.8) (17.3)* (6.3, 7.9)	7.1
458-478	1981	(6.7, 6.3)	6.5
471-491	1980	*(509)* (-) *(17.3)* (7.0, 7.2) (5.6, 5.1)	6.2
491-511	1980	(6.1, 6.0)	6.6
597-614	1977	*(251, 222)* (-) *(13.4)* (6.5, 5.9) (5.4, 6.4)	6.0
614-631	1977	(3.8, 6.5)	5.7
668-688	1975	*(95, 82)* (5.0) (6.0, 7.4, 7.3)	6.0
850-867	1969	*(462)* (-) *(96) (32, 35)* (5.5, 4.7)	5.1
867-884	1969	(4.2, 3.4)	3.8
990-1010	1966	*(114, 123)* (2.3) (4.0, 3.2)	3.0
1010-1030	1965	*(122, 108)* (-) (2.6, 2.8)	2.7

(a) Depths for samples collected in the pits measured at the center of the containers. Depths of the core samples refer to the final 20-cm segments analysed after decontamination.

(b) Estimated by seasonal variation of H_2O_2[19] (pit samples) or $\delta^{18}O$ [21] (core samples); S= Summer; W= Winter.

(c) From Scarponi et al.[15]

(d) In parentheses, data ordered from outside to inside layers of snow cores.

(e) In italics, data from the external, contaminated layers not considered for the mean computation.

(f) Contaminated sample.

(g) Sample not available for Pb determination (see the text).

(h) Adjacent segments of a core section, common data for external and intermediate layers.

ylene speed lathe and successive veneers of snow are chiseled from outside to inside using clean chisels and collecting samples in clean polyethylene bottles. In this case five concentric sub-samples were obtained: the external veneer and three intermediate layers (thickness about 2.0 cm) and the inner core (radius 2.4 cm). The first intermediate layer was reserved for other measurements, thus a four-data radial concentration profile was obtained for each section (see below Table I). In the case of sections with lengths of at least 40 cm, two adjacent segments were chiseled together in a single procedure. This operation led to common samples of the external and intermediate layers, and two separate central cores. Segments of 20-cm length were generally obtained for analysis also in the chiseling procedure, with the only exception of two adjacent segments (17 cm) obtained from a section of insufficient length.

Analytical Instrumentation

The voltammetric system used to perform the lead measurements comprised a PTFE electrochemical cell especially designed[24] for ultratrace metal determinations (Model Rotel 2) and a Polarographic Analyzer (Model 384B), both from EG&G Instruments (Princeton, NJ, USA). The working electrode was a thin mercury film (TMFE) plated on a Rotating Glassy Carbon Disk Electrode (RGCDE). A platinum wire auxiliary electrode and an Ag/AgCl,KCl(sat.) reference electrode were used, both inserted in 2 mm i.d. FEP tubes, filled with saturated KCl/AgCl solution and plugged with porous Vycor tips. The voltammetric instrumentation was installed in a clean chemistry laboratory, under a laminar flow area.

A Perkin-Elmer graphite furnace atomic absorption spectrometer Model 5000 was used for the aluminium concentration measurements.

Lead Measurements

Lead was measured by DPASV. This technique involves, for each determination, the preparation of a freshly formed thin mercury film electrode (TMFE) and the direct voltammetric analysis of the sample solution without any external preconcentration step. The procedure guarantees sufficient sensitivity and a minimum risk of sample contamination.

The TMFE is obtained by controlled potential electrolysis according to the following procedure. The glassy carbon electrode (RGCDE) is prepared by polishing the surface for 10–20 s using a filter paper and wetted γ-alumina (0.075 μm grain size) while the rotation rate is set to 1000 rpm. After repeated washings with ultrapure water and 1:200 diluted HCl, the plating solution (50 g ultrapure water, 200–300 μl ultrapure saturated KCl, 100 μl $Hg(NO_3)_2$ 2.5×10^{-2} M) is put into the electrochemical cell, purged with nitrogen flow for at least 20 min and the electrolytic deposition is carried out at a potential of -1.000 V, while the electrode is rotating at 4000 rpm. After a 20-min plating time rotation is stopped, a 30-s rest time is allowed to pass, the differential pulse potential scan is carried out in the positive direction (scan rate 10 mV s^{-1}, pulse height 50 mV, pulse frequency 5 s^{-1}, final potential -0.180 V) and the voltammogram is recorded. If the voltammogram shows a regular background line, a sufficiently low current (300–500 nA) and no peaks testifying for contamination, the analysis of the sample is begun, otherwise the film is destroyed using the polishing procedure and it is plated again.

The analysis is then carried out as follows. To about 50 g of melted snow sample (exact amount measured at the end of the analysis) 10–20 μl of ultrapure HCl are added as supporting electrolyte and the solution is purged with nitrogen flow (20 min at least). Electrode rotation is switched on (4000 rpm) and the metal deposition is carried out at a potential of between -0.900 to -0.930 V for 45 min. The electrode rotation is stopped and, after a 30-s equilibration time, the stripping voltammetric scan starts and the voltammogram is recorded. The following instrumental settings are used: scan rate 10 mV s^{-1}, pulse amplitude 50 mV, pulse frequency 5 s^{-1}, final potential -0.230 V. At the end of the scan the electrode is conditioned (to remove amalgamated metal completely) holding it at -0.200 V for 5 min with rotation at 4000 rpm. For quantification three subsequent spikes of the metal standard solution (10–20 μl each) are added and the voltammetric measurement is repeated after each addition. Finally the mass of the sample is precisely measured by weighing and a new polishing step is carried out to prepare the glassy carbon electrode for a new determination.

Typical voltammograms obtained for lead determination in a melted snow sample are shown in Figure 2 while the variation of lead peak current against deposition potential (the pseudopolarogram) is shown in Figure 3, where, for comparison, data obtained using KCl as supporting electrolyte are also reported. The half-wave potentials were -0.578 and -0.666 V, respectively in the HCl and KCl supporting electrolytes. As clearly appears from Figure 3, a better defined, steeper pseudopolarographic wave is obtained in the HCl solution, so this electrolyte was used throughout this work. Moreover the deposition potential applied (i.e. from -0.900 to -0.930 V) was selected to assure plateau conditions, but also to allow simultaneous determination of cadmium and lead. Linearity of the lead calibration plot was verified up to about 100 pg/g. In this work only lead results are considered.

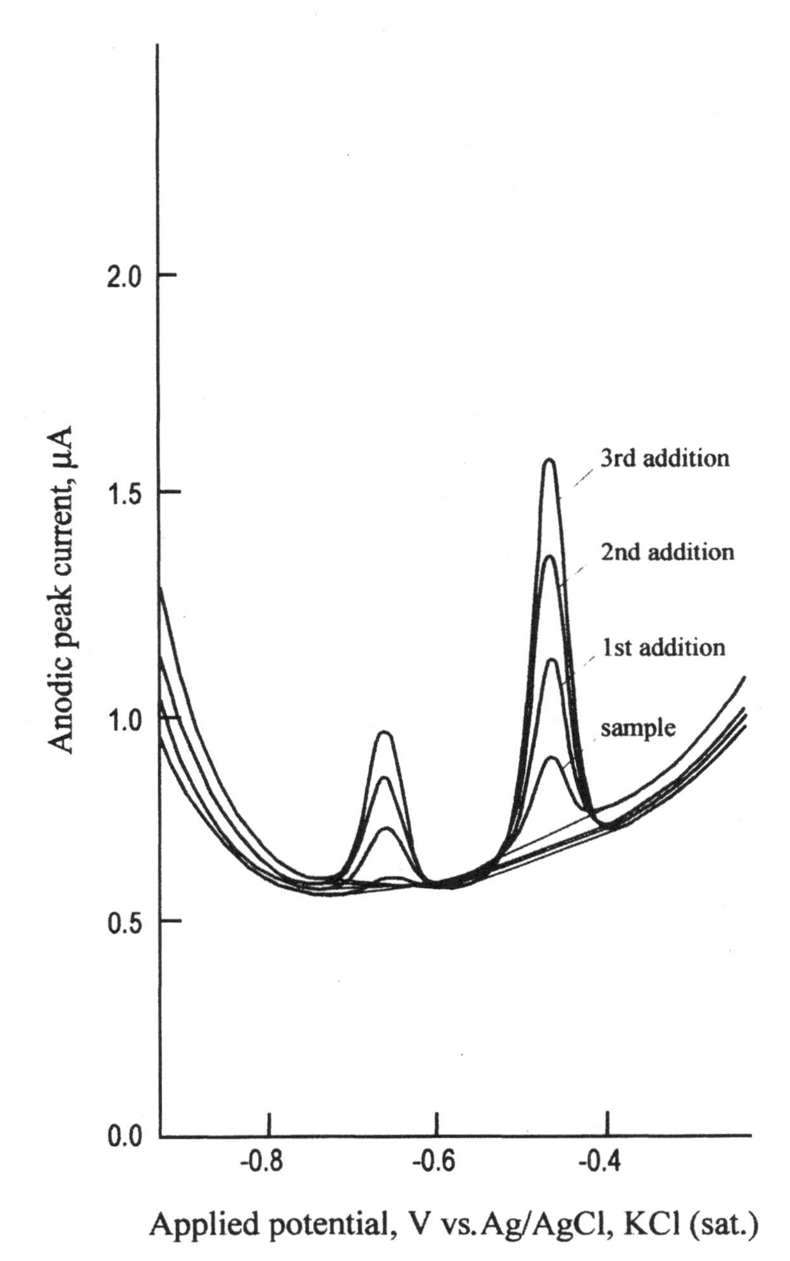

FIGURE 2 Voltammograms obtained in the analysis of one sample of the Styx Glacier plateau (depth 219–239 cm, central core). Additions: 20 μl Pb standard solution (10 μg/l); 10 μl Cd standard solution (20 μg/l).

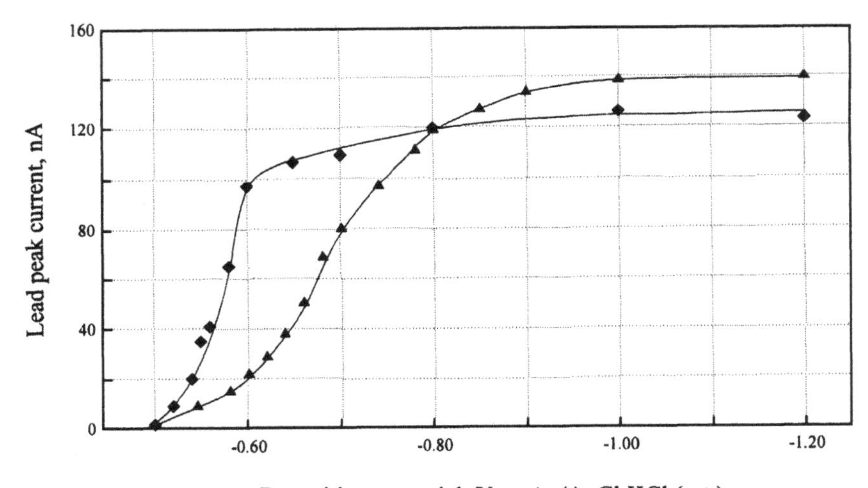

FIGURE 3 Pseudopolarograms of lead obtained from 50 g of melted snow to which 10 μl of HCl NIST (◆) or 20 μl of KCl saturated solution (▲) were added. Deposition time 40 min.

Analytical quality control of lead measurements was routinely carried out analysing the NASS-3 and NASS-4 seawater reference material for trace metals[18]. This material was selected because of the lack of one specific for snow and/or ice. Figure 4 shows a typical control chart obtained during the period of measurements carried out in this work. The mean values of all valid data obtained in the period of measurements, i.e. 39 (SD 5) and 13 (SD 2) ng/l, for

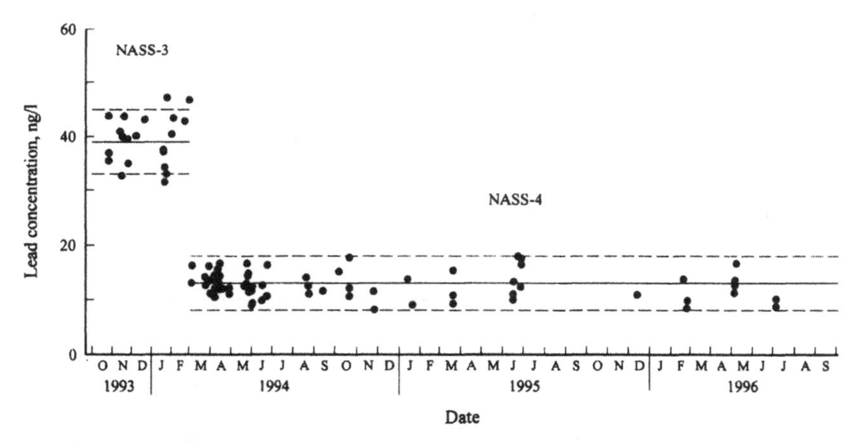

FIGURE 4 Control chart for lead measurements on NASS-3 and NASS-4[18] open ocean seawater reference material. (——) Mean, (----) 95% tolerance limit.

NASS-3 and NASS-4 respectively, are in good agreement with the certified values (see Figure 4). Routinely, if the measured value falls outside the 95% tolerance interval, analytical determinations are interrupted and a special check of the instrumentation is carried out with preparation of a new standard. It is to be noted that accuracy and precision of measurements in melted snow are expected to be better than in the more complex matrix of the reference sample.

No blank correction was applied to instrumental results owing to the ultra clean conditions adopted in all the analytical steps and the particular care taken for contamination control. A negligible blank contribution for the added acid was already experimentally verified[15] in accordance with the certified purity of the acid[17] and with literature data[25]. Containers used for pit sample collection were tested before shipping to Antarctica, verifying that the lead content of ultrapure water stored in them for few weeks was not significanly different from the blank value of fresh ultrapure water. PTFE cell cups and the electrode assembly of the electrochemical cell were periodically tested analysing acidified ultrapure water. Moreover a rough, but sensitive check of the comprehensive blank from the electrochemical cell is obtained routinely before each measurement in the context of the TMFE preparation and control (see above).

As regards precision, considering that the analysis of samples by DPASV is time- and sample-consuming, a pooled estimate of the relative standard deviation was calculated using repetition data obtained on samples collected in the pits. The RSD value was 21%.

Aluminium Measurements

Aluminium measurements were carried out by GFAAS at a wave length of 309.3 nm following the operating procedures described in the literature[26]. 20 μl of sample were used for each determination and a three point calibration curve was used for the quantification.

RESULTS AND DISCUSSION

Changes in Lead Concentration During the Period 1965–1991

Table I reports the lead concentration results obtained in this work for the snow samples collected at the two sites of Victoria Land, together with previously reported, preliminary data[15]. In the case of snow cores, the radial concentration profile is reported, but only reliable data from the inner parts, are used to compute the mean sample concentration. In all cases but three consistent plateau

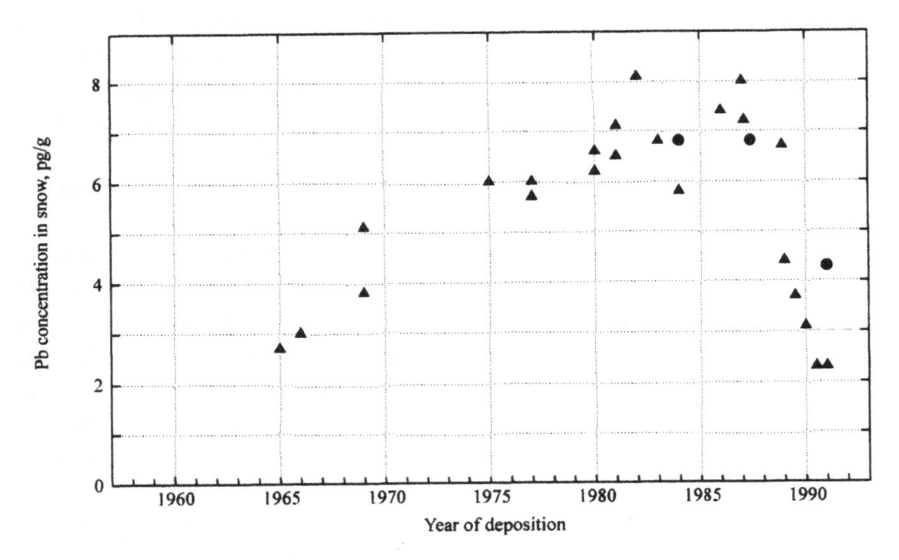

FIGURE 5 Temporal trend of lead concentration in Antarctic snow from Victoria Land for the period 1965–1991. (●) McCarthy Ridge, (▲) Styx Glacier plateau.

values from two inner parts were obtained. This result indicates that contamination is approximately restricted to the external half of snow cores; moreover the radial concentration profiles (see Table I) show no apparent trend of external contamination along the 11-m core of the Styx Glacier plateau. As reported in the table only three core samples are considered contaminated, possibly during the collection and storage processes. The corresponding values are not considered for data interpretation.

The observed Pb concentration range (2.3–8.1 pg/g) appears in good agreement with reliable literature data referred to recent Antarctic snow, as verified at different sites (various sites in East Antarctica, 2.3–7.4 pg/g[10]; Dolleman Island, 1.0–8.8 pg/g[27]; Stake D55, 1.3–8.3 pg/g[12]; Coats Land, 0.8–9.0 pg/g[14]; values given as upper limits, and high values attributed to local contamination, not considered here).

Our data plotted against the year of deposition (Figure 5) show a very interesting pattern of Pb concentration in the period from 1965 to 1991. After a continuous increasing behaviour observed up to the mid-1980s, from 2.7 pg/g to a maximum of 8.1 pg/g, a net decreasing trend is observed down to 4.3 and 2.3 pg/g in McCarthy Ridge and Styx Glacier plateau, respectively, in 1991.

No clear seasonal structure is visible in the collected data (particularly from the Styx Glacier pit) possibly due to the very discontinuous sampling adopted, and the long period of time spanned by each sample collected.

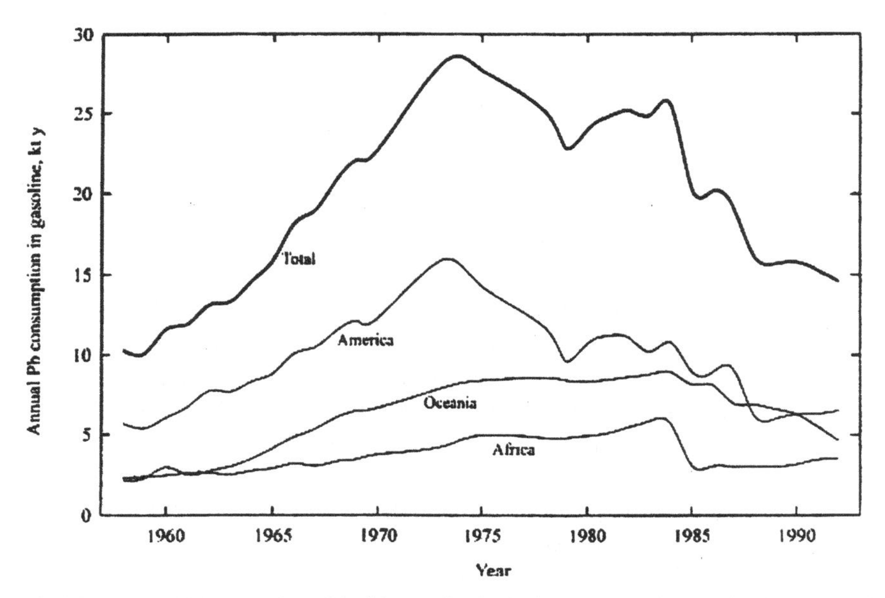

FIGURE 6 Annual consumption of lead in gasoline in the Southern Hemisphere Continents (south of Equator countries). Period 1958–1992. Data from Octel[38] and Fragomeni[39].

The present results confirm and extend previous initial evidence[14–16] concerning the reversing trend occurring in the lead concentration of Antarctic snow within the last two decades. Moreover, considering also all the available data from the literature[12,14–16] one can conclude that the recent historical pattern of lead content in Antarctic snow, with a maximum located between the mid-1970s and mid-1980s, is quite similar, even if in a lower range of concentrations, to that already found in the Arctic (Central Greenland)[5]. Further discussion of the historical pattern in Antarctica will be found in the section on geographical origin (see below).

Natural Contributions

Present-day anthropic contribution to lead concentration in Antarctic snow superimposes on the natural background level arising from rock and soil wind-borne dust, volcanoes, sea-salt spray, forest fires, continental and marine biogenic particulates and volatiles. A rough estimation of the different natural contributions to the total lead concentration has been obtained using the global average of Pb/reference element mass ratios as known for the various sources, though there remains considerable uncertainty and a large variability. Similar procedures had already been attempted by other researchers in the field[10,12,14,27,28].

Lead contribution from rock and soil dust was estimated from the average Al concentration measured in five of the samples analysed for lead, i.e. 0.68 ng/g (SD 0.37 ng/g). Combining this value with the Pb/Al mass ratio in bulk crustal material (9.5×10^{-5})[29], an average contribution of 0.065 pg/g of lead was calculated.

Lead input from volcanic sources was estimated from the Pb/S mass ratio in volcanic emissions ($\sim 7 \times 10^{-5}$[30,31]) and from sulphate concentration measured in samples collected in the same pits dug at the two sites[19], considering that sulphur originating from volcanic emissions has been estimated to be about 12% of S in non-sea-salt sulphate (excess sulphate) in the global atmosphere[31]. Using the mean values of the excess sulphate data[19], i.e. 82 ng/g for McCarthy Ridge and 36 ng/g for the Styx Glacier plateau, average volcanic contributions for lead of 0.2 and 0.1 pg/g were obtained for the two sites.

Lead contribution originating from sea-salt spray was estimated from marine Na concentration in snow and the Pb/Na mass ratio in the sea-derived aerosol which is obtained from the Pb/Na ratio in seawater (i.e. 5.8×10^{-10}: average Pb in Terra Nova Bay, 6.4 ng/l[32], Na in seawater, 11.015 g/l[33]) multiplied by the enrichment factor for sea-derived aerosol relative to bulk seawater (i.e. 100[28,34]). Marine Na was obtained from chloride measurements carried out in the same sites (average chloride concentrations: 1250 and 670 ng/g for McCarthy Ridge and Styx Glacier plateau, respectively[19]), using the Cl/Na mass ratio in seawater of 1.80[33] and considering that chloride in snow of Antarctic costal areas is mainly of marine origin. Contributions of lead from marine aerosol were calculated as 0.04 and 0.02 pg/g for McCarthy Ridge and Styx Glacier plateau, respectively.

From our data it is not possible to calculate the other natural contributions coming from forest fires, continental and marine biogenic particulates and volatiles, but looking at the corresponding fluxes to the atmosphere compared to those of the other sources considered above[30], they are certainly negligible.

In conclusion our estimates led to a total natural contribution of 0.2–0.3 pg/g of lead, with a large margin of uncertainty. However the present result is comparable with other estimates reported in the literature, i.e. 0.25–0.47 pg/g for various sites of East Antarctica[10] and <0.8 pg/g for Coats Land[14].

Anthropic Contributions: Indication of a "Gasoline" Contribution

Comparing the lead concentration observed in our samples with the above estimation for the overall natural contribution, a significant excess ranging from about 2 to 7.8 pg/g is obtained, which in principle can be attributed to local and/or remote anthropogenic sources.

Local contamination, mainly imputable to research stations, has been proved to account for only a few per cent of the total fall-out of lead to the Antarctic continent, and to be restricted to areas very close to the stations[35]. Other local human activities, such as logistic and scientific flights can contribute to a wider dispersion of lead-rich aerosols, but these contributions remain very difficult to estimate. For such reasons, considering also that in the Victoria Land area there are no scientific stations other than the small Italian one, which has only been operating, on a relatively small scale, since 1985, we can reasonably exclude any significant local anthropic contribution to the lead content in snow.

The remote anthropic contribution can be attributed to aerosols originating mainly from southern hemisphere countries, as interhemispheric exchanges are practically prevented[36]. Major sources can be identified with modern non-ferrous metal production processes and the use of leaded gasolines.

Table II reports statistical data for both non-ferrous metal production[37] and the consumption of lead as a gasoline additive[38,39] in southern hemisphere countries. Data refer to selected years of the past three decades and countries are grouped according to the relevant continents. From these data the corresponding lead emissions to the atmosphere are estimated for each source using the average values of emission factors reported in the literature[40,41]. Slight reductions in emission from non-ferrous metal production processes which may have occurred during the last decade are not considered in the computations.

Note that although the amount of lead and other metals involved in non-ferrous metal production is of about two orders of magnitude higher than the lead amount used in gasoline, the corresponding lead emissions are highly in favor of the latter source, due to its considerably higher emission factor. Indeed modern car engines emit, as volatile halogeno-lead compounds, about 75% of the lead present in gasoline[41].

Finally, even if the real end-effect onto Antarctic snow should account for a, possibly different, long-range transport coefficient for the emitted lead by the two types of sources considered, the overall balance at the emission level shows a clear prevalence of the gasoline lead source, at least for the period of time considered here.

A crucial clue to identification of the lead source which caused the characteristic behaviour here observed in Antarctic snow, lies in consideration and comparison of the temporal trends of the two kinds of emissions.

Data reported in Table II show that in this period lead emissions from metal production processes underwent a slight, fairly continuous increase, while emissions from gasoline consumption expanded rapidly during the 1960s and early 1970s, after which there was a net decrease due to policy initiatives taken in

TABLE II Lead sources from non-ferrous metal production and Pb consumption in gasoline from the Southern Hemisphere continents. Annual production/consumption and estimated Pb emissions into the atmosphere for selected years from 1962 to 1992.

Source	Metal production[a] or consumption[b] ($kt\ y^{-1}$)					Pb emitted[c] ($kt\ y^{-1}$)				
	1962	1970	1978	1985	1992	1962	1970	1978	1985	1992
Pb mining										
America S.E.[d]	210.7	227.8	237.4	256	216	0.16	0.17	0.18	0.19	0.16
Africa S.E.	151.3	174.7	133.6	162	102	0.11	0.13	0.10	0.12	0.08
Oceania	376	459.4	418.8	498	575	0.28	0.34	0.31	0.37	0.43
Total	738	861.9	789.8	916	893	0.55	0.65	0.59	0.69	0.67
Pb production										
America S.E.	106.5	130.0	154.9	193.8	176	0.59	0.72	0.85	1.07	0.97
Africa S.E.	82.7	112.7	75.8	81.3	64	0.45	0.62	0.42	0.45	0.35
Oceania	208.1	214.2	243.2	226.1	232	1.14	1.18	1.34	1.24	1.28
Total	397.3	456.9	473.9	501.2	472	2.19	2.51	2.61	2.76	2.60
Cu in Cu-Ni production										
America S.E.	710.3	830.8	1246.4	1335.9	1669	1.39	1.62	2.43	2.61	3.25
Africa S.E.	921.6	1268.2	1326.8	1245.5	771	1.80	2.47	2.59	2.43	1.50
Oceania	87.8	111.6	167.8	167.7	314	0.17	0.22	0.33	0.33	0.61
Total	1719.7	2210.6	2741.0	2749.1	2754	3.35	4.31	5.34	5.36	5.37
Zn in Zn-Cd production										
America S.E.	60.2	105.3	148.1	311.7	339	0.11	0.19	0.27	0.58	0.63
Africa S.E.	136.9	144.1	146.0	180.8	109	0.25	0.27	0.27	0.33	0.20
Oceania	170.6	260.6	290.1	288.7	332	0.32	0.48	0.54	0.53	0.61
Total	367.7	510.0	584.2	781.2	780	0.68	0.94	1.08	1.45	1.44
Total sum[e]						6.77	8.41	9.62	10.25	10.08

TABLE II *continued*

Source	*Metal production[a] or consumption[b] (kt y^{-1})*					*Pb emitted[c] (kt y^{-1})*				
	1962	*1970*	*1978*	*1985*	*1992*	*1962*	*1970*	*1978*	*1985*	*1992*
					Pb from gasoline consumption					
America S.E.	7.75	12.33	11.74	8.98	6.53	5.81	9.25	8.81	6.74	4.90
Africa S.E.	2.57	3.60	4.72	2.95	3.43	1.92	2.70	3.54	2.22	2.57
Oceania	2.75	6.71	8.60	8.16	4.63	2.07	5.03	6.45	6.12	3.47
Total	13.07	22.64	25.07	20.09	14.59	9.80	16.98	18.80	15.07	10.94

a) From United Nations Statistical Yearbook[37].
b) From Octel[38] and Fragomeni[39].
c) Emission data for non-ferrous metal production calculated from average values of emission factor ranges reported by Nriagu et al.[40] (in g t^{-1} metal produced): Pb mining, 750; Pb production, 5500; Cu-Ni production, 1950; Zn-Cd production, 1850. Emission data for gasoline calculated from the fraction of Pb emitted of 0.75 g g^{-1} metal consumed[41].
d) S.E. = south of Equator.
e) For Pb mining and production of metals displayed.

some southern hemisphere countries to limit the emission of the toxic metal into the atmosphere.

The complete data set of annual consumption of lead in gasoline is displayed in Figure 6 and comparison with Figure 5 shows a close relationship between lead concentration in Antarctic snow and lead consumption in gasoline (corr 0.57, $p = 0.02$) when annual mean Pb concentrations are used. Conversely no correlation is obtained when the total sum of lead emissions from the non-ferrous metal production is considered. These results represent a further contribution in favour of the hypothesis[13,14] that most of the lead present in recent Antarctic snow originates from the consumption of leaded gasoline and confirms, as already verified in the northern hemisphere[5–9], that initiatives taken to lower the emission of lead into the atmosphere, are also reflected in a marked decrease of lead spread in the Antarctic.

Geographical Origin

The problem of the geographical provenance of aerosols reaching Antarctica for both present and past climatic conditions has been already debated with contrasting results[13,42–47]. Here an attempt to identify the geographical origin of the lead detected in our samples has been carried out by comparing separately the temporal trend reported in the literature for Coats Land (Atlantic sector of East Antarctica)[14] and that here obtained for Victoria Land (Pacific sector) respectively with the gasoline lead consumption data in South America and in Oceania (Figure 7). Consistent literature data from Adelie Land (Pacific sector)[12] are also displayed on the same figure.

This comparison shows, in the first case, that the lead concentration decrease in Coats Land started in 1975, approximately corresponding to the sharp reduction of lead use in Brazil's gasoline, which had previously contributed highly to the total lead emission of South America[39]. This behaviour reflects a fairly good relationship between lead content in the snow of Coats Land (annual mean values) and lead consumption in the gasoline of the Southern Hemisphere countries of South America (corr 0.47, $p = 0.01$). Conversely the lead concentration trend in Victoria Land seems to reflect more closely the variation in gasoline lead consumption in Australia and New Zealand, which showed a rapid increase from 1960 to 1975, a sort of plateau in 1975–1985, with a maximum in 1984, and then continuously decreased. In fact correlation between the lead data of the present work (annual mean values) and gasoline lead consumption using only data from Oceania is significantly improved (corr 0.80, $p = 0.0002$) with respect to the value obtained above using all Southern Hemisphere data (corr 0.57, $p = 0.02$).

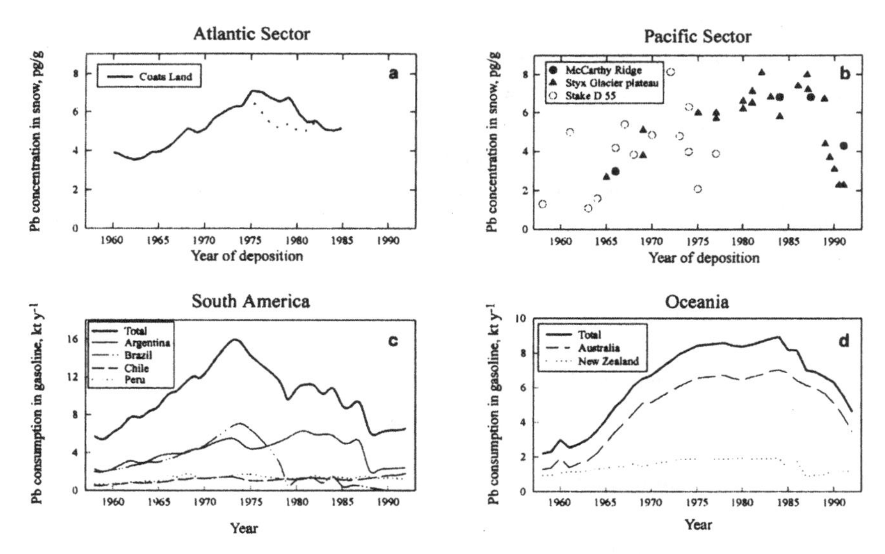

FIGURE 7 Comparison of the temporal trends of lead concentration in the snow of (a) the Atlantic sector of Antarctica (Coats Land[14]), and (b) the Pacific sector (Victoria Land, present data, and Adelie Land, Stake D 55[12]), respectively with the gasoline lead consumption in (c) South America and (d) Oceania[38,39].

These different behaviours suggest that, under the present-day climatic conditions, South America and Oceania are the predominant source areas for aerosols reaching Atlantic and Pacific sectors of Antarctica, respectively. It is expected that more defined and detailed patterns will soon be available for lead in snow in the two Antarctic sectors. However, further, possibly conclusive, contributions with respect to the present hypothesis could be obtained from new data from, e.g., isotopic measurements and rare earth element determinations or other tracers.

CONCLUSIONS

DPASV confirmed its potential in terms of the extremely high sensitivity which allows the direct determination of lead in Antarctic snow, down to the low pg/g level, to be carried out without any sample preconcentration. This feature, together with the easy transportability of the instrumentation, is continuously exploited by us to make on-site measurements during Italian expeditions to Antarctica, where a clean chemistry laboratory is available at the permanent summer station of Terra Nova Bay.

The new data obtained in this work confirm that lead concentration in the snow of Victoria Land decreased markedly during the second half of the 1980s. The concomitant reduction of lead consumption as gasoline additive in Southern Hemisphere countries, particularly Australia and New Zealand, led us to hypothesize that this source was prevailing in the past three decades and that the origin of aerosols reaching the Pacific sector of Antarctica is presumably Oceania.

Acknowledgements

This study was performed in the framework of Projects on "Environmental Contamination" and "Glaciology and Paleoclimatology" of the Italian *Programma Nazionale di Ricerche in Antartide* and financially supported by ENEA through co-operation agreements with the Universities of Venice and Milan, respectively.

We are very grateful to all the members of Italian technical staff for the excellent logistic support they gave and to the researchers of the Analytical Chemistry group led by one of the authors (G.S.) for the skilful sampling activity. Special thanks are due to M. Frezzotti for his precious help in the choice of sampling sites, to G. Rampazzo for his assistance during the GFAAS measurements, to P. Cescon for useful discussions during the preparation of the manuscript, to C. F. Boutron (Laboratoire de Glaciologie et Géophisique de l'Environnement du CNRS, Grenoble, France) who patiently reviewed the paper and to V. M. Thomas (Center for Energy and Environmental Studies, Princeton University, Princeton, NJ) for the useful exchange of lead consumption data.

References

[1] V. Thomas and T. Spiro, In: *Industrial ecology and global change* (R. Socolow, C. Andrews, F. Berkhout and V. Thomas, eds., Cambridge University Press, Cambridge, UK, 1994) pp. 297–318.

[2] C.F. Boutron, *Environ. Rev.*, **3**, 1–28 (1995).

[3] H. Oeschger and C.C. Langway Jr, Eds. *The environmental record in glaciers and ice sheets* (Wiley, Chichester, UK, 1989) pp. 400.

[4] S. Hong, J.P. Candelone, C.C. Patterson and C.F. Boutron, *Science*, **265**, 1841–1843 (1994).

[5] C.F. Boutron, U. Görlach, J.P. Candelone, M.A. Bolshov and R.J. Delmas, *Nature*, **353**, 153–156 (1991).

[6] K.J.R. Rosman, W. Chisholm, C.F. Boutron, J.P. Candelone and S. Hong, *Geochim. Cosmochim. Acta*, **58**, 3265–3269 (1994).

[7] R. Lobinski, C.F. Boutron, J.P. Candelone, S. Hong, J. Szpunar–Lobinska and F.C. Adams, *Environ. Sci. Technol.*, **28**, 1459–1466 (1994).

[8] R. Lobinski, C.F. Boutron, J.P. Candelone, S. Hong, J. Szpunar–Lobinska and F.C. Adams, *Environ. Sci. Technol.*, **28**, 1467–1471 (1994).

[9] J.P. Candelone, S. Hong, C. Pellone and C.F. Boutron, *J. Geophys. Res.*, **100**, 16605–16616 (1995).
[10] C.F. Boutron and C.C. Patterson, *J. Geophys. Res.*, **92**, 8454–8464 (1987).
[11] C.F. Boutron and C.C. Patterson, *Geochim. Cosmochim. Acta*, **47**, 1355–1368 (1983).
[12] U. Görlach and C.F. Boutron, *J. Atmos. Chem.*, **14**, 205–222 (1992).
[13] K.J.R. Rosman, W. Chisholm, C.F. Boutron, J.P. Candelone and C.C. Patterson, *Geophys. Res. Lett.*, **21**, 2669–2672 (1994).
[14] E.W. Wolff and E.D. Suttie, *Geophys. Res. Lett.*, **21**, 781–784 (1994).
[15] G. Scarponi, C. Barbante and P. Cescon, *Analusis*, **22**, M47–M50 (1994).
[16] G. Scarponi, C. Barbante and C. Turetta, *Terra Antartica Reports*, **1**, 103–106 (1997).
[17] P.J. Paulsen, E.S. Beary, D.S. Bushee and J.R. Moody, *Anal. Chem.*, **60**, 971–975 (1988).
[18] NASS-3, NASS-4, Open Ocean Seawater Reference Material for Trace Metals (1990, 1992). National Research Council of Canada, MACSP, Ottawa, Canada K1A 0R6.
[19] G. Piccardi, R. Udisti and F. Casella, *Intern. J. Environ. Anal. Chem.*, **55**, 219–234 (1994).
[20] A. Neftel, P. Jacob and D. Klockow, *Nature*, **311**, 43–45 (1984).
[21] E. Barbolani, M. Dini, M. Frerrotti, V. Maggi, G. Piccardi, F. Serra, B. Stenni and R. Udisti, *Terra Antartica Reports*, **1**, 65–70 (1997).
[22] G. Piccardi, E. Barbolani, S. Becagli, R. Traversi and R. Udisti, In: *Proceedings of 4th National Congress on Environmental Contamination* (Venice, December 6–7 1995). pp. 53–62.
[23] J.P. Candelone, S. Hong and C.F. Boutron, *Anal. Chim. Acta*, **299**, 9–16 (1994).
[24] L. Mart, H.W. Nürnberg and P. Valenta, *Fresenius Z. Anal. Chem.*, **300**, 350–362 (1980).
[25] E.D. Suttie and E.W. Wolff, *Atmos. Environ.*, **27A**, 1833–1841 (1993).
[26] C.F. Boutron and S. Martin, *Anal. Chem.*, **51**, 140–145 (1979).
[27] E.D. Suttie and E.W. Wolff, *Tellus*, **44B**, 351–357 (1992).
[28] C.F. Boutron and C.C. Patterson, *Nature*, **323**, 222–225 (1986).
[29] S.R. Taylor and S.M. McLennan *The continental crust: its composition and evolution. An examination of the geochemical record preserved in sedimentary rocks* (Blackwell Scientific Publications, Oxford, UK, 1985), 300 pp.
[30] J.O. Nriagu, *Nature*, **338**, 47–49 (1989).
[31] C.C. Patterson and D.M. Settle, *Geochim. Cosmochim. Acta*, **51**, 675–681 (1987); Erratum, **52**, 245 (1988).
[32] G. Scarponi, G. Capodaglio, G. Toscano, C. Barbante and P. Cescon, *Microchem. J.*, **51**, 214–230 (1995).
[33] T.R.S. Wilson, In: *Marine electrochemistry* (M. Whitfield and D. Jagner, eds., Wiley, Chichester, UK, 1981) pp. 145–185.
[34] C.P. Weisel, R.A. Duce, J.L. Fasching and R.W. Heaton, *J. Geophys. Res.*, **89**, 11607–11618 (1984).
[35] C.F. Boutron and E.W. Wolff, *Atmos. Environ.*, **23**, 1669–1675 (1989).
[36] D. Wagenbach, In: *Chemical exchange between the atmosphere and polar snow* (E. W. Wolff and R. C. Bales, eds., Springer, Berlin, 1996) pp. 173–199.
[37] U.N. (1965–1992). *United Nations Statistical Yearbook* (U.N., New York).
[38] Octel (1984–1994). *Worldwide Gasoline Survey*. The Associated Octel Company Ltd., London, U.K., various issues.
[39] A.S. Fragomeni, *Lead consumption in gasoline in Brazil*. Petrobras Research and Development Centre, Rio de Janeiro, Brazil. Personal communication (1995).
[40] J.O. Nriagu and J.M. Pacyna, *Nature*, **333**, 134–139 (1988).
[41] P. Falchi, D.A. Gidlow and G.R. Searle, *How to obtain rapid improvements in air quality*. (Società Italiana Additivi per Carburanti s.r.l., Milan, 1992).
[42] S. Joussaume and J. Jouzel, *J. Geophys. Res.*, **98**, 2767–2805 (1993).
[43] F.E. Grousset, P.E. Biscaye, M. Revel, J.R. Petit, K. Pye, S. Joussaume and J. Jouzel, *Earth Planet. Sci. Lett.*, **111**, 175–182 (1992).
[44] A. Gaudichet, M. De Angelis, S. Joussaume, J.R. Petit, Y.S. Korotkevitch and V.N. Petrov, *J. Atmosph. Chem.*, **14**, 129–142 (1992).
[45] M. De Angelis, N.I. Barkov and V.N. Petrov, *Nature*, **325**, 318–321 (1987).
[46] M. De Angelis, N.I. Barkov and V.N. Petrov, *J. Atmosph. Chem.*, **14**, 233–244 (1992).
[47] R.J. Delmas and J.R. Petit, *Geophys. Res. Lett.*, **21**, 879–882 (1994).

MULTIPARAMETRIC APPROACH FOR CHEMICAL DATING OF SNOW LAYERS FROM ANTARCTICA

R. UDISTI

Department of Public Health and Environmental Analytical Chemistry, University of Florence, 9 Via Gino Capponi, I-50121 Florence, Italy

A dating method for successive snow layers is proposed which is based on a combination of concentration profiles of three chemical parameters measured for each sample: H_2O_2, MSA, and $nssSO_4^{2-}$. In the studied area (Northern Victoria Land, Antarctica), these substances demonstrate a clear seasonal character with summer maxima and winter minima which together can constitute an univocal annual indicator. The proposed method involves searching for maximum values and normalizing the concentration/depth profile of each substance; a smoothed sum of the contribution for each component, for each depth value, gives a resulting profile which is better adapted to objective interpretation of the seasonal trends.

This method is applied to the dating of snow and firn samples coming from two snowpits and one shallow firn core at three different stations, which are found at different altitudes and distances from the sea, within approximately 200 Km from the Italian base at Terra Nova Bay.

The variations in concentration of the three substances with depth were examined and the relative trends were evaluated as a function of the geographic position of the sampling stations.

KEY WORDS: Snow layer dating, seasonal trends, multiparametric dating method, annual indicators, snow analysis, antarctica.

INTRODUCTION

The recognition of successive annual snow layers is an important instrument for the relative dating of snow, firn, or ice samples taken from snowpits or firn/ice cores. By counting successive annual layers, with reference to the surface or to a known temporal event found in a particular annual layer, it is possible to determine the absolute date for each snow layer sampled. This dating, and even more so the seasonal setting of the analysed sample, is useful in many areas of study, including:

* the study of seasonal trends of various components;
* the identification of principal and secondary sources of the components;
* the identification and study of transport mechanisms;
* the interpretation of existing correlations between the various components;
* the characterization of a station from the perspective of average annual accumulation;
* the importance of altitude, distance from the sea and geographic position on snow fall accumulation.

Generally, relative dating of successive snow layers is carried out following the seasonal trend of only one parameter, such as H_2O_2, NO_3^-, Na^+, Cl^-, Cl^-/Na^+, $\delta^{18}O$, non sea salt sulphate ($nssSO_4^{2-}$), methanesulphonic acid (MSA), electrical conductivity (ECM), stratigraphy or by "visual" comparison of concentration profiles with the depth for more than one parameter[1–19]. These methods, however, can be difficult to interpret objectively. Therefore, a method to obtain a dating hypothesis based on less subjective criteria seems to be very useful.

In the present work a simple multiparametric method of dating is proposed which is based on a linear combination of concentration of H_2O_2, MSA and $nssSO_4^{2-}$, measured for the same sample of snow or firn, after a process of normalization and smoothing carried out on concentration profiles with the depth for each of the three components.

EXPERIMENTAL

Instrumentation

The determination of H_2O_2 is carried out using a method of Flow Injection Analysis with a spectrofluorimetric detector (Shimadzu RF 551). This method is based on the formation of a fluorescent dimer for the reaction of p-hydroxy-phenyl-acetic acid with H_2O_2 in the presence of peroxidase[20,21].

The determination of MSA and of the ionic species necessary for the determination of the concentration of $nssSO_4^{2-}$ (total SO_4^{2-}, Na^+ or Cl^-) is carried out by ionic chromatography using a Dionex 4000i Ion Chromatograph with anionic or cationic conductivity suppressor and Dionex AI-450 intergration software.

Further details regarding methods used, manipulation of samples, contamination problems, and analytical results obtained are described in our previously published works[22–25] and other work in preparation[26]. Table 1 shows a synthesis of the determination methods and the principal operative variables.

Reagents

In ion chromatographic measurements, for the preparation of eluents, regenerants and all standard solutions, ultrapure water (> 18 mΩ) was used, obtained from a Millipore MilliQ apparatus and continuously recycled in an Elga UHQ apparatus. The water produced was then bidistilled further, using a solution of $KMnO_4$ in the first phase of distillation to destroy trace residues of H_2O_2, and then used for the H_2O_2 determination[26]. Stock standard solutions (1000 mg/l) were obtained from Merck Suprapur, when available, or Merck, Fluka or Sigma Reagent Grade reagents.

Sampling

The proposed method was applied to samples taken from two snowpits of approximately 2 m excavated during the Italian Antarctic Expedition of 1990/91 and from a shallow firn core, approximately 7.5 m deep, obtained during the Italian Antarctic Expedition of 1992/93. The Italian Antarctic Expeditions were a part of the Antarctic National Research Program (PNRA). The three sampling stations, located in Northern Victoria Land (Antarctica) and within a range of 200 Km from the Italian base at Terra Nova Bay,

Table 1 Methods for concentration determination of the compounds utilised as seasonal indicators.

H_2O_2

Reagents: 3.9 10^{-2} M p-hydroxy-phenyl-acetic acid
0.13 M NH_3
8.4 10^{-4} M EDTA
375 Peroxidase units/100 ml
Carrier: H_2O Milli-Q bidistilled on $KMnO_4$ and filtered on 0.45 µm.

MSA

Separator Column: Dionex AS5A-5 µm (guard column: Dionex AG5A)
Eluent: 1 ml/min 1.8–30 mM $Na_2B_4O_7$
Conductivity suppressor: Dionex AMMS-1
Regenerant: 7.0 ml/min 1.25 10^{-2} M H_2SO_4

nss-SO_4^{2-}

SO_4^{2-} and Cl^-
Separator Column: Dionex AS4A (guard column: Dionex AG4A)
Eluent: 2 ml/min 1.2 10^{-3} M $NaHCO_3$ + 1.3 10^{-3} M Na_2CO_3
Conductivity suppressor: Dionex AMMS-1
Regenerant: 3.0 ml/min 1.25 10^{-2} M H_2SO_4

Na^+

Separator Column: Dionex CS 10 (guard column: Dionex CG10)
Eluent: 1 ml/min 2 10^{-2} M HCl + 5 10^{-4} M Diaminopropionic acid (DAP)
Conductivity suppressor: Dionex CMMS-2
Regenerant: 2.0 ml/min 5 10^{-2} M Tetrabutyl-ammonium hydroxide (TBAOH)

differ in altitude and distance from the sea. This difference is useful for the evaluation of average annual accumulations in the studied area and the understanding of phenomena related to the origin and transport of substances present in the atmospheric aerosol, which make up a part of the snow composition[4,10,25,27–36]. The geographic position and characteristic of the sampling stations are reported in Figure 1 and Table 2.

Snowpit samples (47 samples from Station 19, 58 samples from Station 27) were collected by inserting pre-cleaned polyethylene vials (35 × 100 mm) into the vertical snow walls after having removed about 10 cm of snow from the exposed surface. Anti-contamination procedures were followed during sampling, storage (at –20°C), and manipulation of the samples.

Results relative to the analysis of a series of smaller vials (16 × 100 mm) taken in parallel columns from the same snowpits have been previously published. The results of the respective concentration trends, as a function of the depth, are in excellent agreement. This is evidenced by the comparison of Figures 2 and 3 of the present work with Figures 1, 2, 7 and 8 of Piccardi *et al.*[25]. This fact is of particular importance with regards to the concentration of $nssSO_4^{2-}$, the determination which carries the greatest risk of experimental error[37]. The findings seem even more significant considering the fact that in the previous series, due to a shortage of sample volume, it was not possible to determine the cations; as a consequence the previously obtained values for $nssSO_4^{2-}$ were determined as a function of Cl^- content. Noting that the relative trends for the two series of vials for each station are superimposable, we can deduce that for stations of this typology and in this area the use of Na^+ or Cl^- as indicators of the sea salt contribution is fundamentally the same. This holds true particularly for stations situated at low elevations and near the sea[27].

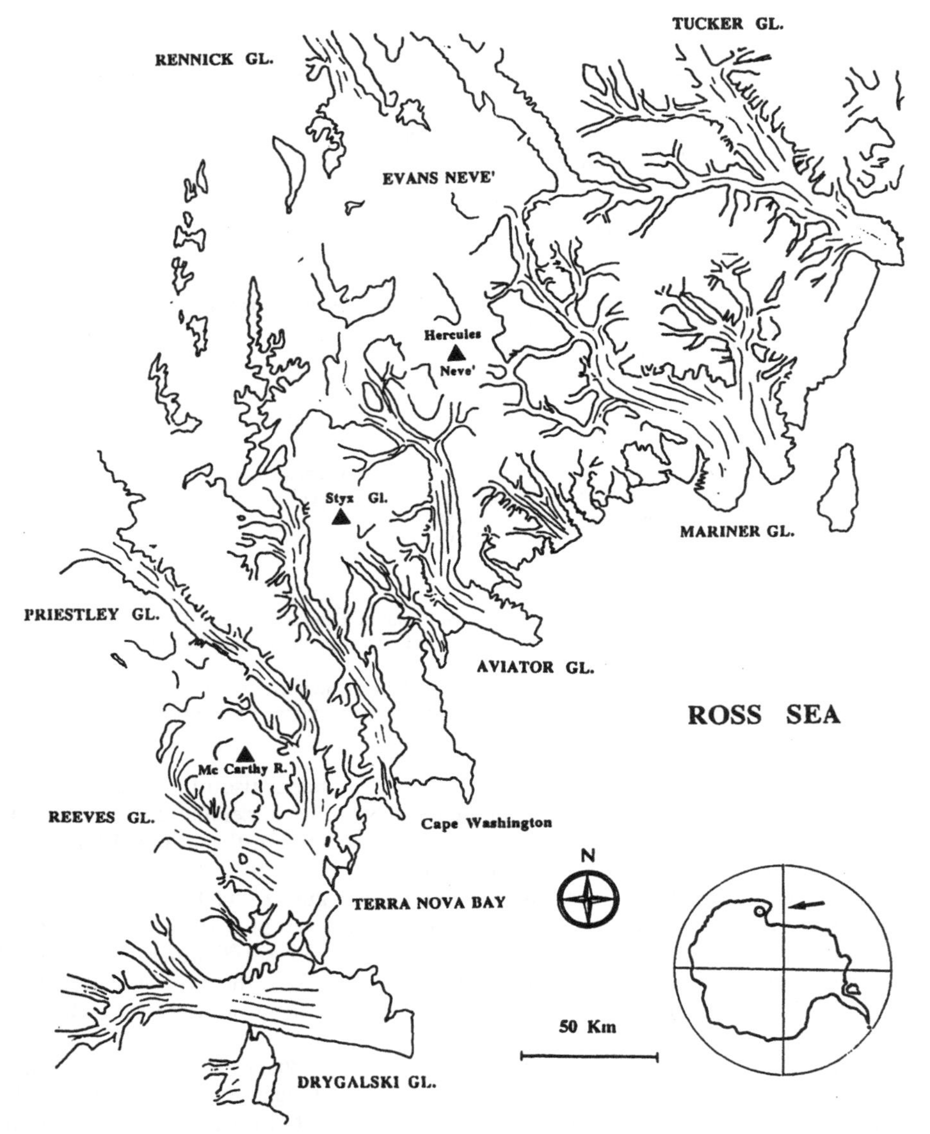

Figure 1 Sampling map of stations sampled during Italian Antarctic Expeditions 1990/91 (Mc Carty Ridge and Styx Glacier) and 1992/93 (Hercules Nevé).

Table 2 Sampling stations.

Station Code	*Station Name*	*Lat. Sud*	*Long. Est*	*Height m a.s.l.*	*Km from coast line*
27 SN IV/B	Mc Carthy Ridge	74°32'57"	162°56'29"	700	40
19 SN IV/B	Styx Glacier Plateau	73°51'54"	163°41'30"	1700	50
36 FC VIII	Hercules Neve'	73°07'34"	164°58'12"	2990	90

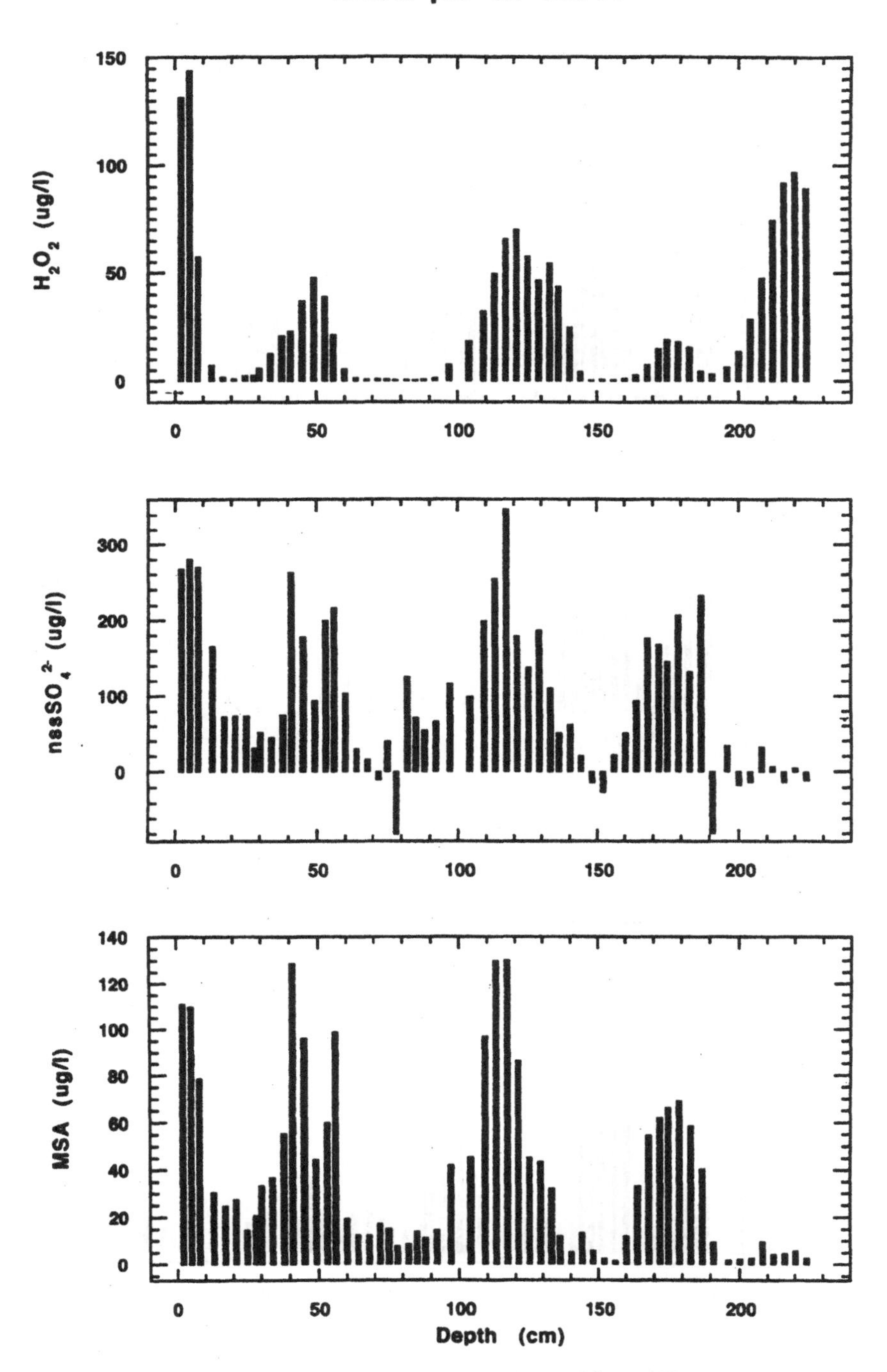

Figure 2 Station 27—Original concentration profiles for H_2O_2, $nssSO_4^{2-}$ and MSA.

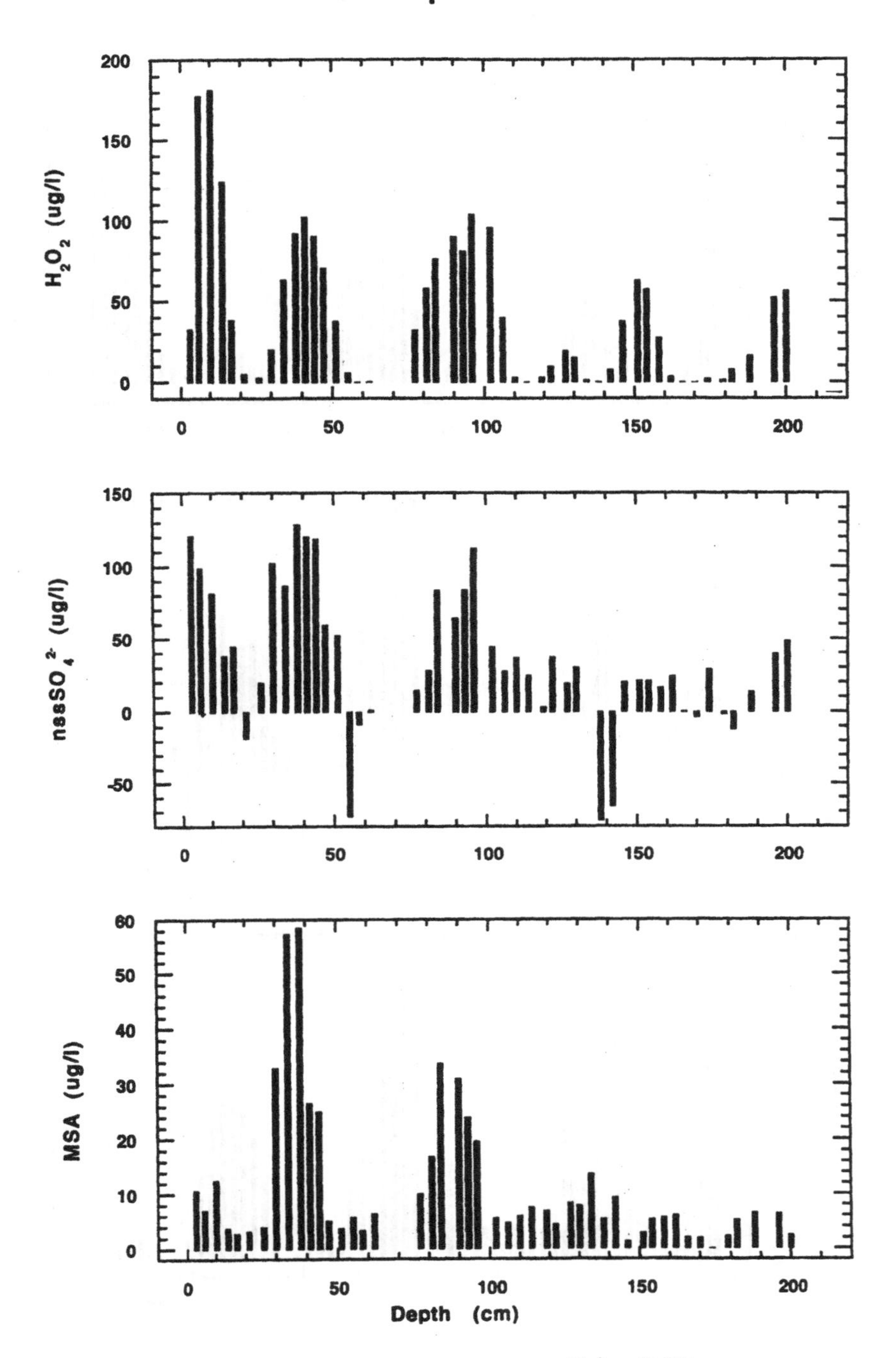

Figure 3 Station 19—Original concentration profiles for H_2O_2, $nssSO_4^{2-}$ and MSA.

The firn core from Station 36 (Hercules Neve') was transported frozen (at –20°C) to Italy in 50–70 cm long pieces in two sealed polyethylene sacks. Working in a cold chamber and under a laminar flow hood, each piece of firn core was cleaned removing approximately 1 cm of the exterior and then divided into sub-samples (total of 234) each about 3 cm long. The sub-samples were kept in pre-cleaned polyethylene containers and placed in double polyethylene sacks. The samples were thawed immediately prior to taking measurements under a class 100 laminar flow hood. Control blanks were tested periodically for the sampling and analysis procedures and for the cleanliness of the containers, which showed contamination values below the detection limit or at least two orders lower than the concentration values determined in the samples.

DISCUSSION

The contemporary consideration of more than one parameter with the aim of dating can present some limits:

1. The choice of parameters to be considered.
   It seems appropriate to choose those chemical substances which present clear and unequivocal seasonal concentration variations, due either to seasonality at the source or in the transport mechanisms. The choice of parameters is also tied to the typology of the sampling station, with particular attention to aerosol of marine origin (primary and secondary) and, therefore, to the station's geographic position.
2. The non-contemporaneousness of concentration maxima relative to the various components in the annual temporal series.
3. The lack, or difficult interpretation, of some concentration maxima for one or more of the observed components.
4. The presence of annual maxima relatively lower than other nearby annual maxima, or in any case belonging to the same temporal series, for a given component.

With regards to the first point, given the goal of obtaining an evaluation of the dating exclusively by chemical means, dating techniques based on the $\delta^{18}O$ isotopic ratio, ECM and stratigraphy have not been included.

Based on points 1 and 2, neither the concentration profile of the Na^+ or Cl^- ions nor the Cl^-/Na^+ ratio have been taken into consideration for dating. In fact the Na^+ ion does not appear to be a reliable seasonal indicator in our study area due to the strong sea influence. As previously mentioned[25,27], the study area is characterized by a relatively high number of precipitation events such as salt storms with notab'y high values of Cl^- and Na^+, which are found at the same ratio in sea water. These precipitation events, more frequently a winter occurrence, are not so rare in summer, or at least, in the seasons of transition, which are characterized by considerable variability in the peri-Antarctic atmospheric circulation.

Neither nitrate concentration seems to be useful for dating in studied area because the seasonal signal is less distinct with increasing depth[25].

On the basis, therefore, of previous analyses of snow from snowpit samples in the Northern Victoria Land, the following substances have been selected as parameters indicating seasonality:

* H_2O_2
* $nssSO_4^{2-}$
* MSA

Examining the seasonal trends of the concentrations of these compounds in the snow or firn samples, also in the case of these parameters, problems due to seasonal maxima which do not perfectly correspond persist. In fact, H_2O_2 presents concentration maxima as a function of the period of maximum solar radiation (late-spring, early-summer), while the other two substances are tied to the development of marine life, particularly to that of algal bloom (late-summer). The dephasing among the parameters will be, therefore, elevated as much as the number and extent of snow deposition occurring in the critical period between the respective maxima (early- to mid-summer). Moreover, this dephasing will depend, considerably, on the sample resolution and, therefore, both on the extent of the annual deposition which characterizes the station and on the sampling technique.

These limitations are, however, unavoidable and can be considered acceptable if they cause a widening of the annual peaks which is not too exalted. If the widening is excessive it can impede the complete distinction between summer maxima and winter minima.

Some fundamental observations regarding the three parameters selected for the dating of collected snow samples should be examined.

H_2O_2

H_2O_2 is principally produced from radical reactions originating from photolysis of O_3[38]. Since this is a photochemical process, maximum concentration of H_2O_2 in the atmosphere is reached during the period of maximum solar radiation, which coincides, in the studied area, with late-spring to early-summer. The H_2O_2 which is contained in snow remains stable for long periods of time[17,39,40]. In areas where the accumulation of snow is undisturbed, not burdened by the possibility of diffusion layers caused by superficial fusion of summertime snow (sufficiently low summer temperatures), the seasonal variation of the H_2O_2 content is one of the most reliable chemical methods for sequential dating of successive snow layers. Numerous Authors[3,8,9,11–13,16,17] have used this parameter for dating snow samples coming from snowpits, firn cores and ice cores of moderate depth. Positive comparisons with the dating method based on isotopic $\delta^{18}O$ ratio variations have permitted the use of this parameter as a reliable seasonal indicator[11].

The determination of H_2O_2 in snow is carried out on the sample as soon as it is thawed because this compound is not stable in aqueous solution, above all not at very low concentration (a few μg/l).

Table 3 reports the obtained values for linearity, reproducibility, and detection limits.

The Figure 4 shows box plots relative to the distribution of H_2O_2 concentration for the three different stations. One can note the progressive increase of median concentration

Table 3 Analytical parameters for spectrofluorimetric (H_2O_2) and ion chromatographic determinations.

	Linearity range *μg/l*	*Sensitivity* *$nS\ \mu g^{-1}\ l$*	*St. deviation* *(5 means.)*	*Detection limit* *ng/l*
H_2O_2	0– 200	–	5%	10
Na^+	0–3000	70.5	2%	155
Cl^-	0–1500	32.5	3%	175
SO_4^{2-}	0–1000	8.2	3%	220
MSA	0– 250	13.1	3%	225

The loops used were 1000 μl for anions and cations and 200 μl for H_2O_2. The standard deviation was calculated for standard solution with concentration about 50 times higher than the detection limit.

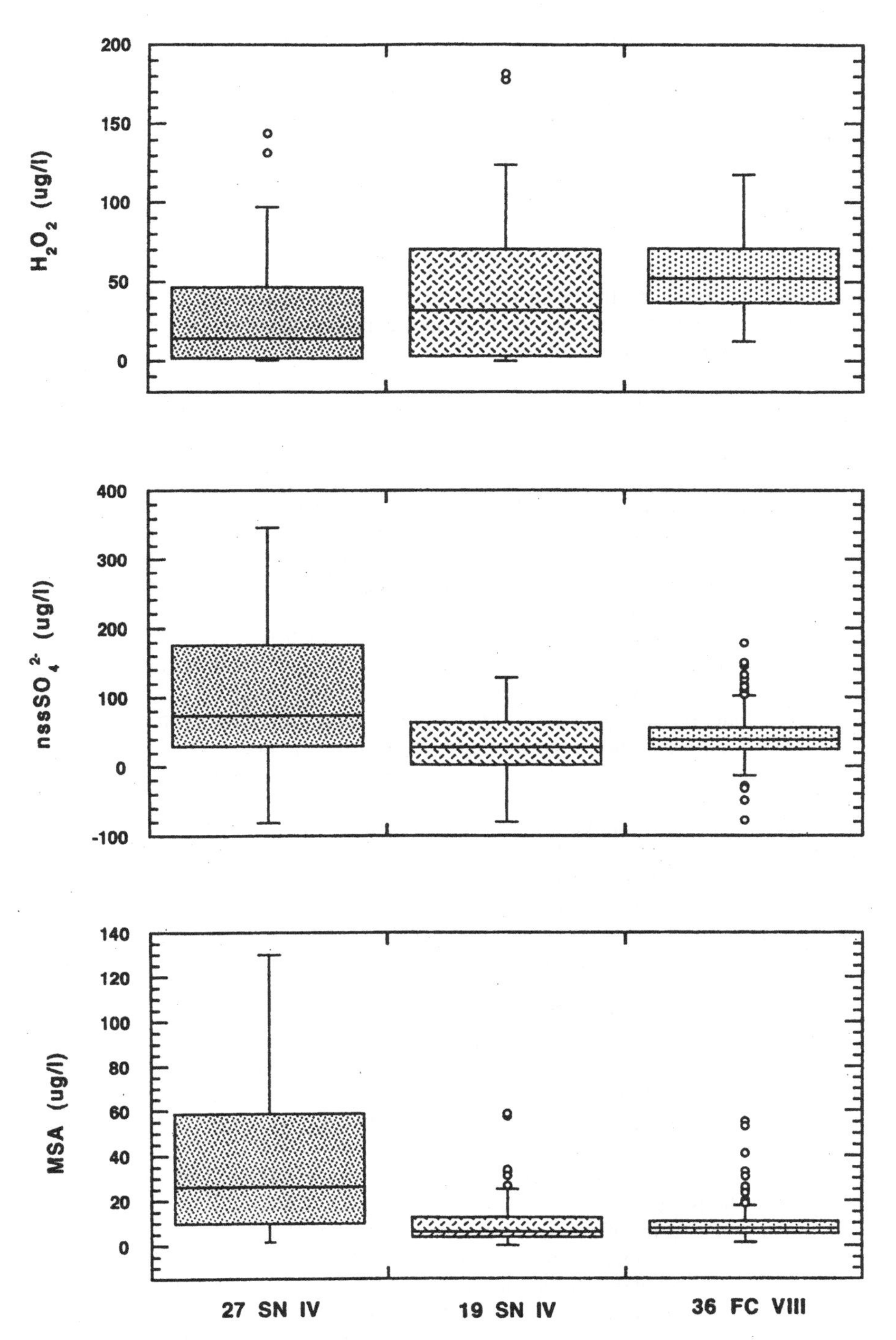

Figure 4 Box plots indicating the concentration distributions of the considered compounds in the three stations.

(line in the box) with altitude. The range encompassing 50% of the samples (box) appears to be sufficiently ample to be able to give a significant seasonal variation between high summer and low winter values. For the highest station (36 FC VIII) the twenty-fifth percentile line (bottom line in the box) is shifted toward values which are clearly higher than the other two stations. This implies the presence of relatively higher winter minima.

nssSO_4^{2-}

The term nssSO_4^{2-} refers to the contribution of sulphates not coming from sea spray to the total sulphate concentration. For each sample considered, the nssSO_4^{2-} concentration is calculated using the following formula:

$$[nssSO_4^{2-}] = [SO_4^{2-}]_{tot} - [SO_4^{2-}]_{sw}$$

where: $[SO_4^{2-}]_{sw} = [X]_s * ([SO_4^{2-}]/[X])_{sw}$

and $[SO_4^{2-}]_{sw}$ is the sample concentration of sulphate from sea spray, $[X]_s$ is the concentration of the indicator element for sea contribution (Na^+ or Cl^-) in the sample, and $([SO_4^{2-}]/[X])_{sw}$ is the ratio between sulphates and Na^+ or Cl^- present in sea water. Therefore, for the nssSO_4^{2-} concentration calculation, all sodium or chlorides found in the sample are considered to be of marine origin.

In Antarctica, the principal contribution to nssSO_4^{2-}, at least during summer, comes from the oxidation of biogenic dimethylsulphide (DMS)[8,41–48] with respect to other postulated sources such as volcanic origin, crustal origin, and long range transport effect[19,42,46,49–53]. Due to this fact, the nssSO_4^{2-} concentration presents maxima in the period of phytoplanktonic bloom and in the immediately following period. For the studied area this period is during the months of January-February (full-late-summer).

The clear seasonality of the nssSO_4^{2-} concentration has been evidenced both by measurements taken from snow in snowpits, ice and firn cores[6,7,18,25,29,34,54] as well as in aerosol measurements[55–59]. The presence, therefore, of summer maxima in the concentration profiles of this component can be utilised for dating purposes.

The level of nssSO_4^{2-} concentration depends largely, in the studied area, on the geographic position and in particular on the distance from the sea and the altitude. Figure 4 shows the distribution of this component for the three different stations examined (box plot). One can note the distinct reduction of the median concentration and the maximum values with altitude passing from 700 m a.s.l. at Station 27 to 1700 m a.s.l. at Station 19. The levels then remain relatively constant for Station 36 (2990 m a.s.l.). In addition, with increasing altitude there is a narrowing of the range which encompasses 50% of the data. This is in accordance with the fact that the principal source is from the sea, even if the distribution can be considered atmospheric, given that the oxidation of DMS into SO_2 takes place in the atmosphere.

The small range between summer maxima and winter minima, however, does not compromise its use as a seasonal indicator, even at higher altitudes.

From an analytical point of view, the measurement of nssSO_4^{2-} concentration depends therefore on the linear combination of two analytical measurements: determination of Na^+ or Cl^- concentration and determination of total SO_4^{2-} concentration in each sample. The determinations are carried out using ionic chromatography. Table 3 reports the most significant analytical characteristics. Since these measurements are obtained by the difference between two experimental data, there is greater possibility of error, compared with those for the determination of MSA and H_2O_2[37,60,61], for the determination of $nssSO_4^{2-}$ concentration. Naturally, to the analytical error the contribution (not easily quantifiable) due to the initial hypothesis must be added: the ratio $([SO_4^{2-}]/[X])_{sw}$ which is known and constant, the concentration of Na^+ and Cl^- in the sample which is attributable only to marine contribution, the absence of selective fractionation phenomena.

For Station 19 and 27 the $nssSO_4^{2-}$ concentration was established by taking the concentration of Na^+ as references for marine origin. For Station 36, measurements for the determination of the cation concentration are still in progress, and therefore the $nssSO_4^{2-}$ concentration values have been taken from the concentration of Cl^-.

MSA

Methanesulphonic acid derives only from oxidative processes of DMS of phytoplanktonic origin (same reference as $nssSO_4^{2-}$; see above). This compound, hence, comes from the same origin and precursor as the most important summer contributor of $nssSO_4^{2-}$, but with the difference that the origin of MSA is univocal and can, therefore, be utilised as a reliable indicator of biological marine activity[23,25,28,29,44,45,62,63]. The seasonal trend of this compound is the same as for the $nssSO_4^{2-}$ (assuming similar atmospheric transport mechanisms) with concentration maxima present in late-summer. Also for this substance a strong seasonality is evidenced both from measurements of the snow cover (snowpit and firn core)[6,34] and from aerosol[55–59,64], and this seasonality has made it possible to use MSA for dating successive snow layers[25,29]. There is, however, a difficulty for use of this parameter as a seasonal indicator. Low concentrations are found, also during summer, in areas far from the sea and at high elevations. This can be noted, in fact, from the distribution of this component for the three stations (Figure 4); the box plots relative to MSA show a distinct lowering of both the median value and, above all, of the summer maxima with increased elevation. In particular, for Station 36, 90% of the samples had a MSA content between 1 and 20 µg/l. Therefore, particularly sensitive methodologies are necessary which are able to detect such low concentrations without turning to pre-concentration techniques, which are normally affected by contamination problems. Table 3 shows the analytical performance of the method used. Notwithstanding the small range of concentration values of MSA for Station 36, the difference between summer and winter concentrations is sufficiently clear to render the concentration vs. depth profile of this component useful for dating.

Dating

Figure 2 reports data relative to the temporal distribution of H_2O_2, $nssSO_4^{2-}$ and MSA concentrations for snowpit 27 SN IV, and likewise for snowpit 19 SN IV in Figure 3. A decisive seasonal behaviour appears evident for all the concentration profiles. One can note the presence of summer peaks which are relatively low in some years and peaks which are very high for others. For example, at a depth of approximately 125 cm for snowpit 19 and 175 cm for snowpit 27, the concentration of H_2O_2 presents a small peak with particularly low values. This peak seems to be, with good probability, a summer maxima since the same situations is found in an analogous way for the two snowpits which were sampled contemporaneously (the fourth peak starting from the surface; presumed date: summer 1987/88).

As for the concentration profiles of the other two components, similar behaviour can be observed but some summer peaks are more pronounced and clearly confirm the H_2O_2 trend, while others are much more ambiguous.

To be able to compare summer maximum peaks on a mathematical basis, even if they do not perfectly coincide on a temporal level, all annual peaks for each series relative to the various components must be "normalized". This term is used to mean the construction of a concentration profile in which peaks, which are with good probability maximum summer indicators, are assigned the same numerical importance independent of their absolute value.

The curve for the concentration profile can, therefore, be expressed as a series of ratios between each concentration value and the nearest maximum within a defined interval in the neighbourhood of that value. Therefore, the maximum value of each peak, if it is sufficiently isolated from the other picks, will take on the numerical value = 1. Thus, all the peaks will assume the same numerical importance, independent of the effective value of the concentration reached for that year for that parameter.

The equation used is the following:

$$\text{n.v.} = \frac{\text{a.v.}}{\max(\text{a.v.} \pm q)}$$

where n.v. = normalized value
a.v. = analytical value

Where max (a.v. ± q) represents the maximum value of concentration found in the interval of amplitude 2 q + 1 (q data before, the chosen value, and q data after) with respect to the analytical value considered. The search for maxima becomes, therefore, a mobile search procedure. This procedure could be defined as normalization by mobile maximum in a defined interval.

Obviously, particular care is given to the choice of interval in the neighbourhood of each experimental value in the search for maxima. In fact, there is risk of considering as summer peaks oscillations of relatively low winter values by using an interval which is too small; in practice each higher value in its small neighbourhood could be recognised as a summer peak and could assume the value = 1. Instead, by using interval values which are too large there is the risk of not recognising, as summer peaks, small peaks which are positioned between two annual peaks presenting much higher maximum values. The ideal situation could be that which permits full recognition of all certainly summer peaks independent of their maximum value. Hence it is clear that the ideal system would involve a prior knowledge of summer peaks, while this knowledge is actually the objective. A compromise could be to set the interval of search for maxima equal to the number of samples between the most distant peaks which can certainly be attributed to the summer period. Thus, the resulting error is almost exclusively conservative in that there is risk of not giving value = 1 to small peaks found between higher peaks. This occurs, however, only in particularly unfavorable situations.

Once a normalized profile is obtained, it is possible to recognize the annual trends more clearly and, by iterative process, to be able to choose a smaller interval. This makes it possible to better highlight peaks found near other higher peaks, which appear to have a fairly clear connotation of summer maximum. On the other hand, it is quite improbable that an unfavorable situation for the recognition of a small peak would exist contemporaneously for all three components in the same year. By summarizing the normalized values of all three components point by point it is possible to obtain a new temporal profile in relation to depth. This parameter sum presents high values (maximum value = 3) for depth where normalized peaks are present for all three examined substances; instead, very low values are relative to snow layers for which there is no peak value. Intermediate peaks are described for layers in which at least two sufficiently high normalized peaks overlap.

Due to the considerable dispersion of the original data and the imperfect contemporaneousness of the annual maxima for the three components, peaks which are somewhat widened and articulated have been obtained making some smoothing necessary. Figures 5 and 6 show normalized profiles for the three parameters and their smoothed (order 3) sums for the two snowpits.

In the case of the two snowpits, the good interpretability of the H_2O_2 concentration profiles permits the selection of a relatively small interval, considerably inferior to the number of samples between the two most distant summer maxima (18 points, Station 27, Figure 5). This interval was established as 11 points; each concentration value was divided by the maximum concentration value measured for the five samples before and the five after the examined sample. The choice of such a limited interval permits easy recognition of the small peaks for H_2O_2 positioned at a depth of approximately 180 cm for snowpit 27 (Figure 2) and approximately 130 cm for snowpit 19 (Figure 3). It can be noted how the normalized concentration profiles better evidence the summer maxima making it easier to base the annual accumulation calculation on them. The sum profiles obtained do not seem to demonstrate ambiguity of interpretation for either station: the annual peaks have been numbered progressively and an absolute dating has been superimposed for the two snow pits (Figures 5 and 6). The numbering of the summer peaks has been based on the normalized sum profile and some peaks can therefore do not appear in the temporal series of the individual compounds (for istance, peak n. 5 for MSA ad $nssSO_4^{2-}$ in the station 27—Figure 5).

It is interesting to note how the profiles of the original data are not well-defined (for example, layers deeper than 100 cm for MSA and $nssSO_4^{2-}$ for Station 19, Figure 3) but are clearer after normalization (peaks, 4, 5, and 6—Figure 6).

The normalized profile of $nssSO_4^{2-}$ for Station 27 (Figure 5) offers a good example of how trends of difficult interpretation, relative to only one of the normalized profiles, are toned down in the summed profile. In fact the presence of three maxima near each other at a depth of 70–140 cm makes peak n. 3 difficult to interpretate for $nssSO_4^{2-}$; since the other two compounds are quite evident at the same depth, that peak becomes clear and not ambiguous in the normalized sum profile.

The annual and average annual accumulation values obtained are in complete agreement with previous reports on annual accumulation for the same snowpits for the series of samples taken in parallel columns[25,30]. The previous annual snow accumulation values were based simply on a visual comparison of the original concentration profile, paying particular attention to the H_2O_2 profile which appears to be the most reliable for the two snow pits.

For Hercules Heve' firn core (samples 36 FC VIII) instead, simple visual observation of the concentration profiles for the three parameters does not permit good discernment of the summer peaks though the H_2O_2 concentration still seems to be the most reliable for most of the depth scale (Figure 7). As already evidenced by the box plot of H_2O_2 for this Station (Figure 4), elevated concentration values even in winter are often observed particularly in the last part of the firn core. This section appears, instead, to be better explained by the temporal progression of the other two parameters which have unclear profiles at other depths (for istance, between 200 and 350 cm). However it is evident that a reliable dating based on clear and univocal seasonal signals is not possible from just a visual analysis of the concentration profiles of each component. The normalization of the concentration peaks for each component (given the very different concentration values

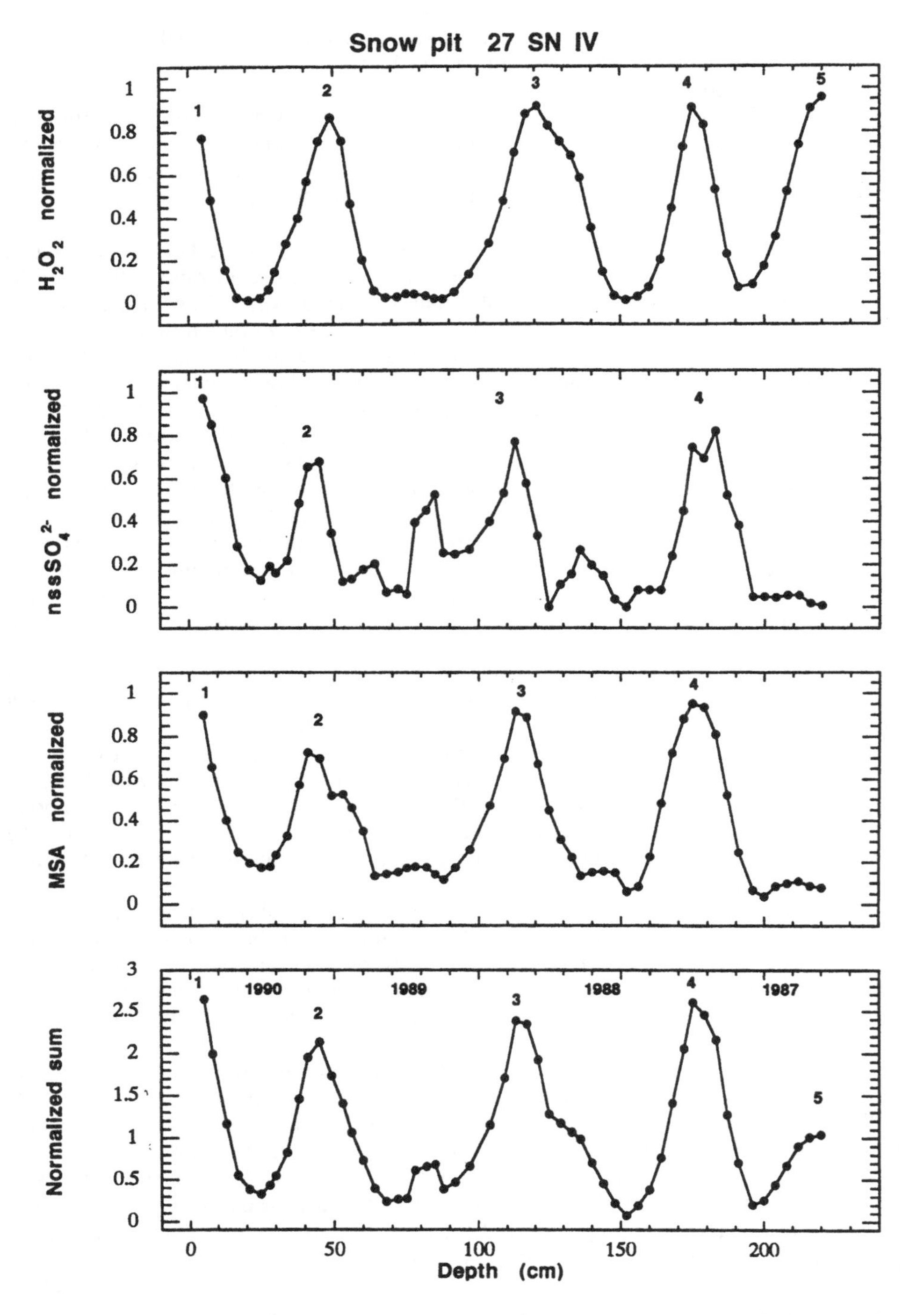

Figure 5 Station 27—Normalized profiles for H_2O_2, $nssSO_4^{2-}$ and MSA concentrations and their sum. The individuation of annual summer peaks is performed on the basis of the normalized sum profile (see text).

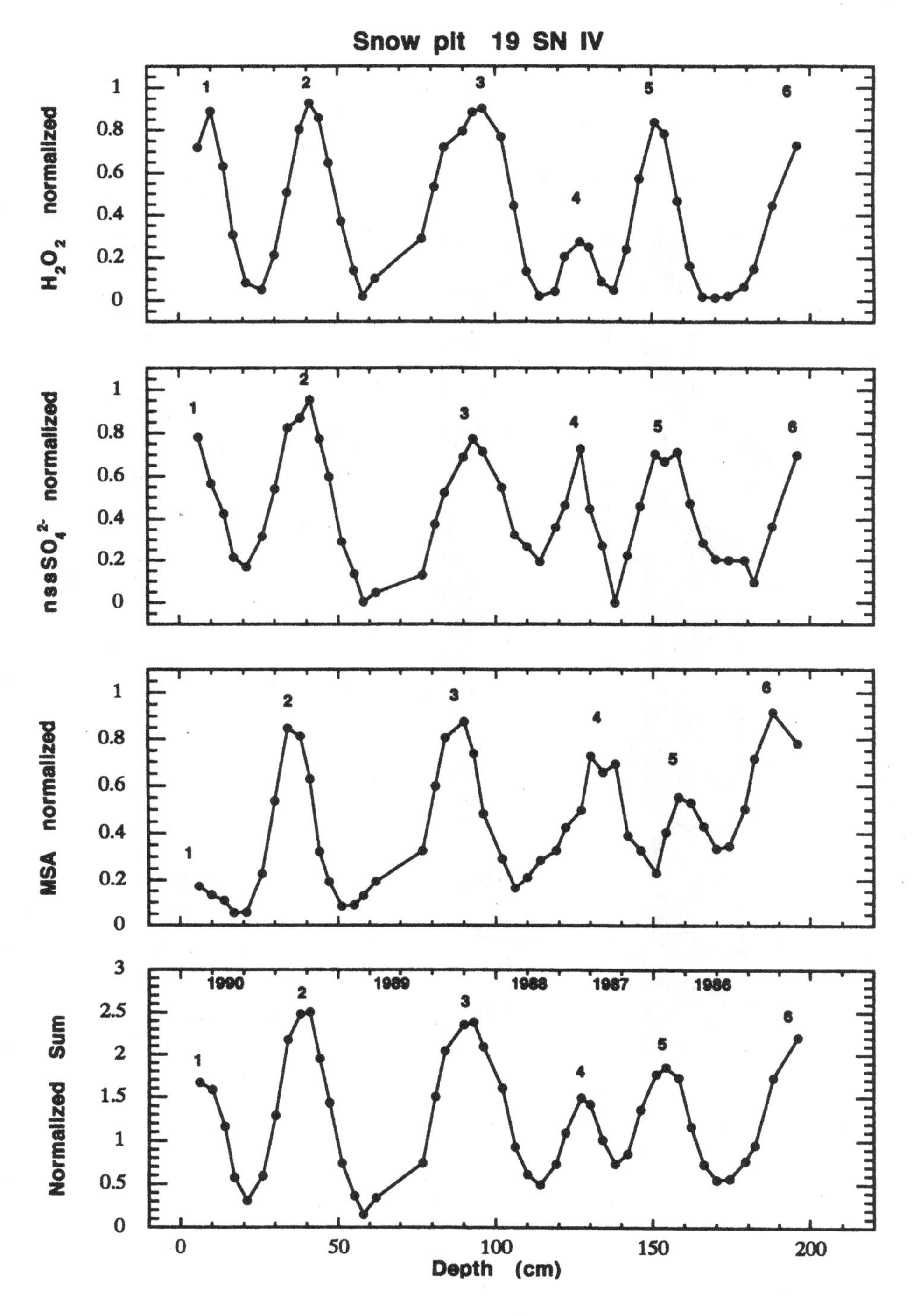

Figure 6 Station 19—Normalized profiles for H_2O_2, $nssSO_4^{2-}$ and MSA concentrations and their sum. The *individuation* of annual summer peaks is performed on the basis of the normalized sum profile (see text).

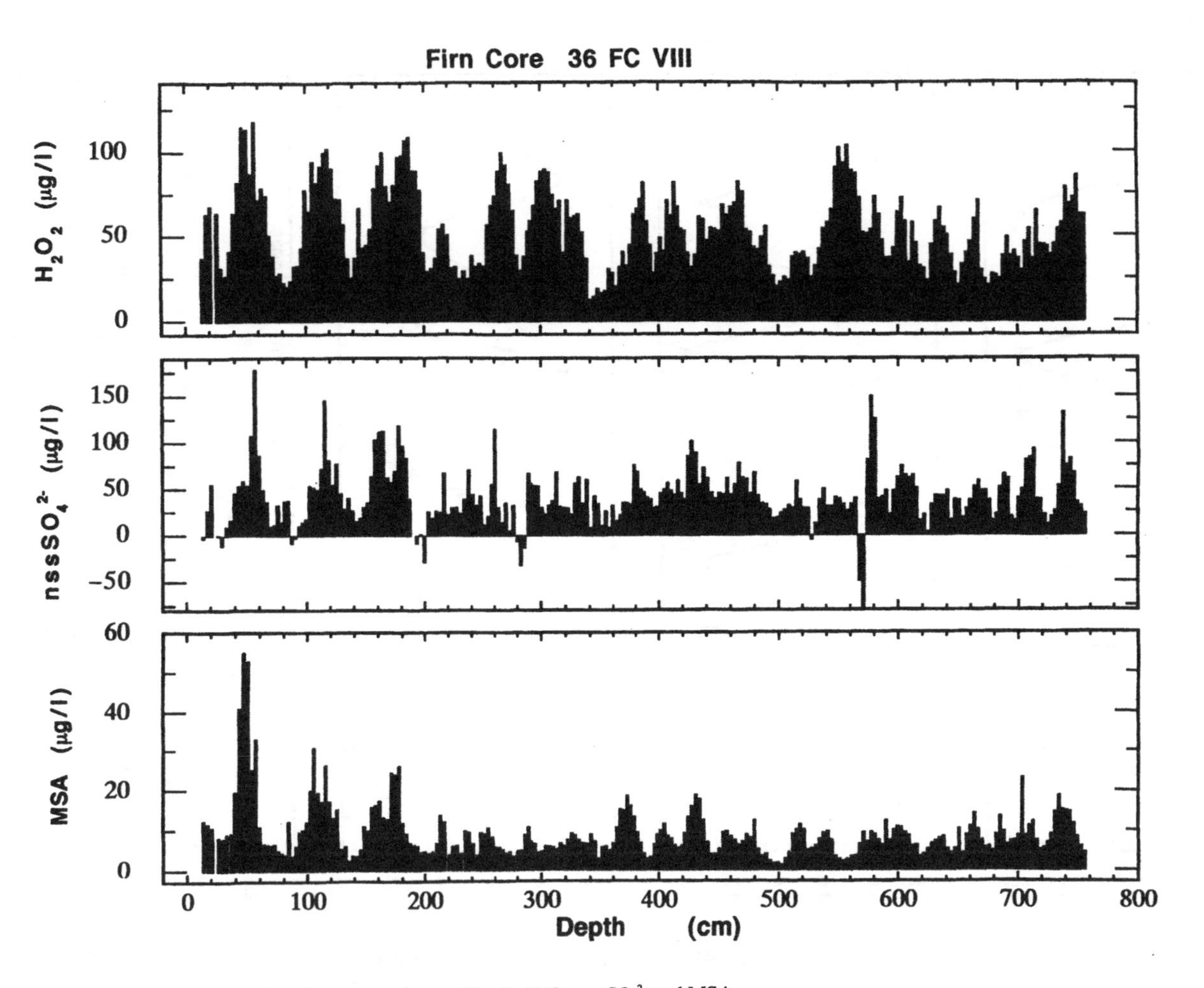

Figure 7 Station 36—Original concentration profiles for H_2O_2, $nssSO_4^{2-}$ and MSA.

that the summer peaks have in every temporal series) and the evaluation of the temporal trend of their sum seems, therefore, inevitable to obtain an acceptable resolution of the summer peaks. Another difficulty is presented by double peaks which can at times be considered distinct or effectively relative to only one year, since they do not appear double in all three observed parameters.

Due to the relatively elevated dispersion of the original concentration data, in the case of the Hercules Neve' samples, a light smoothing (order 3) of the data before calculating the normalized values was preferred, so as to obtain curves which were more easily interpretable. In Figure 8 the results of the three normalizations and the representative sum for the temporal trends for the normalized parameters are reported. Here an interval of 21 values was chosen for the search for maxima. This interval is prudently more ample compared to that for the two snowpits due to the increased difficulty in interpreting the seasonal trends for each component. The supposed summer indicator peaks have also been numbered progressively on the basis of the normalized profile sum. Different peaks that might belong to the same year have been indicated with the same number (e.g. 4 and 4').

By examining the result it is possible to highlight summer maxima which are generally well-defined. The fact that they are relative to three different parameters and obtained with different methodologies reinforces the probability of their existence as true summer peaks.

The normalized profiles of $nssSO_4^{2-}$ and MSA are well-defined even at depths of 200–350 cm where original profile concentrations cannot be reliably evaluated. Even though the presence of multiple peaks makes not clear the profile interpretation in this range of depth, the normalized profiles of the two substances are very similar to each other, especially the two groups of peaks numbered 5, 5' 6 and 7, 7', 8. In general, the normalized temporal series for $nssSO_4^{2-}$ and MSA are quite similar for the entire depth range. The normalized H_2O_2 values are instead different from the other two components but it is precisely this difference in the data series which confirms or eliminates some poorly resolved peaks as annual indicators.

An excellent agreement between the three normalized profiles is noted for peaks 1, 2, 3, 4, 4' 6, 9, 11, 13, 17 and 20. Relatively small or poorly defined peaks in the H_2O_2 profile are fully confirmed by the other two profiles (peaks 8, 10, 12) or vice versa (peaks 9', 14, 15, 16). Peaks 18 and 19 do not appear in the H_2O_2 profile but they are very distinct and reliable in the MSA and $nssSO_4^{2-}$ profiles. Peaks 5 and 7 which are single for H_2O_2 and double for MSA and $nssSO_4^{2-}$ cannot be clearly or certainly interpreted.

By analyzing the sum profile it is possible to almost certainly identify 20 annual peaks plus 4 others that can be attributed to distinct years or to the same year (peaks 4', 5' 7' and 9'). In conclusion, it is possible to individuate 22 ± 2 summer peaks, with a 10% margin of uncertainty for the entire firn core. This margin is acceptable for the purpose of estimating single annual and average annual accumulations for this firn core.

The resulting water equivalent annual average deposition of about 160 Kg/m^2 Yr is in good agreement with that reported by Allen III (1985)[2] for a near area (182 Kg/m^2 Yr—core E10, Rennick Glacier area).

Integration of the available data with those still being obtained from the approximately 10 m firn cores, taken at various stations during the last Antarctic campaign (1993/94), can result in a valid discussion about the reliability of dating methods and a qualitative and quantitative characterization of snowfall in Northern Victoria Land.

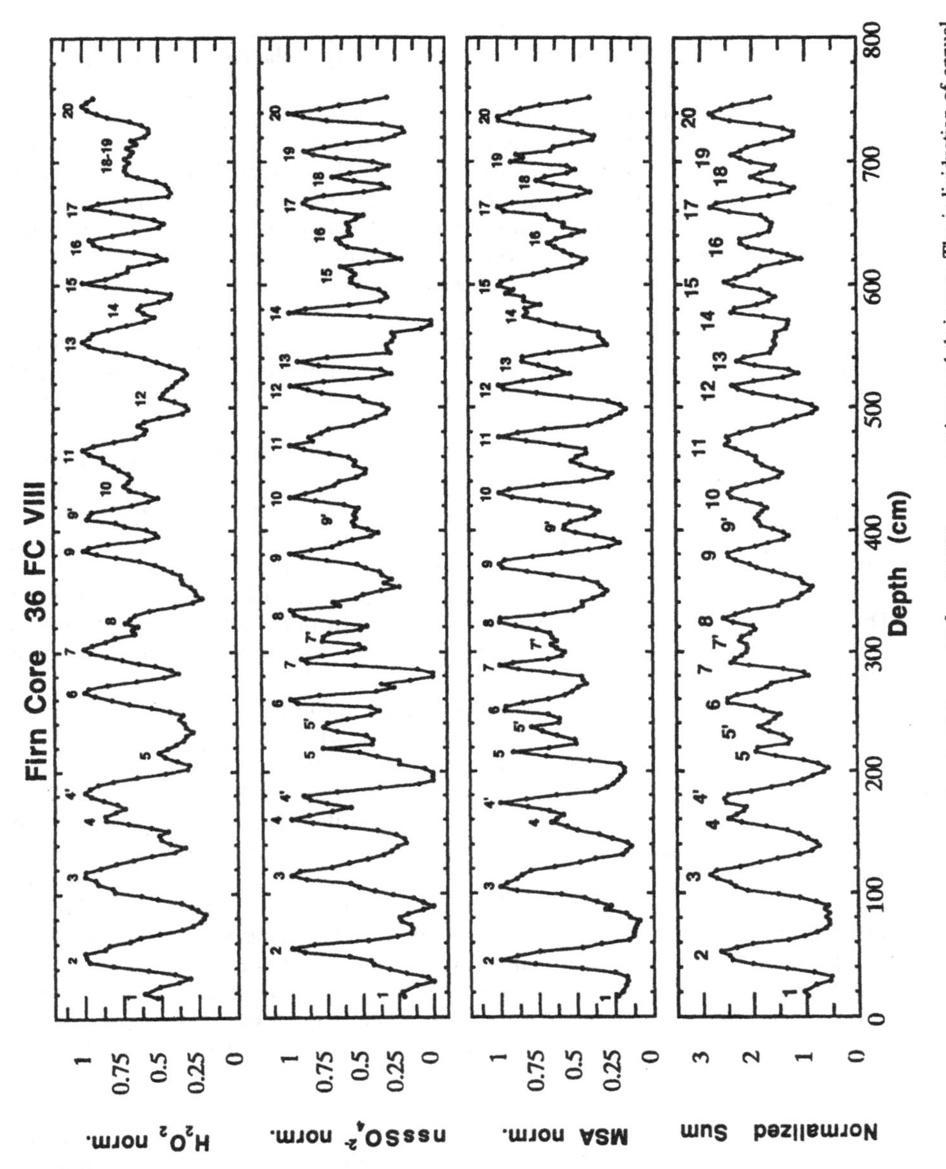

Figure 8 Station 36—Normalized profiles for H_2O_2, $nssSO_4^{2-}$ and MSA concentrations and their sum. The individuation of annual summer peaks is performed on the basis of the normalized sum profile (see text).

CONCLUSIONS

It is possible to obtain a more reliable estimation of annual snow layers using a sum of three normalized profiles of seasonal indicators, rather than observing original concentration trends.

The sum of normalized values gives maximum values which are sufficiently defined and reliable with greater sampling frequency and less temporal dephasing of the measured parameters.

The mobile search for defined interval maxima and normalization of concentration values appears easily applicable and seems to give satisfactory results, provided that an optimal interval is selected using iterative processes.

The obtained results are in agreement with the trend of average accumulations for Northern Victoria Land as a function of altitude and distance from the sea.

Acknowledgement

Research carried out in the framework of a Project on Glaciology and Paleoclimatology of the *Programma Nazionale di Ricerche in Antartide,* and financially supported by ENEA through a cooperation agreement with *Universita'degli Studi di Milano.*

References

1. M. De Angelis and M. Legrand, *J. Geophys. Res.*, **99–D1**, 1157–1172 (1994).
2. B. Allen III, P. A. Mayewski, W. B. Lyons and M. J. Spençer, *Ann. Glaciol.*, **7**, 1–6 (1985).
3. S. Whitlow, P. A. Mayewsky and J. E. Dibb, *Atmos. Environ.*, **26A**, 2045–2054.
4. P. A. Mayewsky, *Ann. Glaciol.*, **14**, 186–190 (1990).
5. P. A. Mayewsky and M. R. Legrand, *Nature*, **346**, 258–260 (1990).
6. C. C. Langway Jr, K. Osada, H. B. Clausen, C. U. Hammer, H. Shoji and A. Mitani, *Tellus*, **46B**, 40–51 (1994).
7. E. Mosley-Thompson, J. Dai, L. G. Thompson, P. M. Grootes, J. K. Arbogast and J. F. Paskievitch, *J. Glaciol.*, **37**, 11–22 (1991).
8. R. J. Delmas, *Sci. Total Environ.*, **143**, 17–30 (1994).
9. V. I. Morgan, I. D. Goodwin, D. M. Etheridge and C. W. Wookey, *Nature*, **354**, 58–60 (1991).
10. E. Isaksson and W. Karlen, *J. Glaciol.*, **40**, 399–409 (1994).
11. A. Sigg and A. Neftel, *Ann. Glaciol.*, **10**, 157–162 (1988).
12. A. Sigg, T. Steffelbach and A. Neftel, *J. Atmos. Chem.*, **14**, 223–232 (1992).
13. A. Sigg, K. Fuhrer, M. Anklin, T. Staffelbach and D. Zurmuhle, *Environ. Sci. Technol.*, **28**, 204–209 (1994).
14. J. Neubauer and K. G. Heumann, *Fresenius Z. Anal. Chem.*, **331**, 170–173 (1988).
15. J. Neubauer and K. G. Heumann, *Atmos. Environ.*, **22**, 537–545 (1988).
16. A. Neftel, P. Jacob and D. Klockow, *Tellus*, **38D**, 262–270 (1986).
17. A. Neftel, in: *NATO ASI Series, Vol.* G28—*Seasonal Snowpacks*, (T. D. Davies *et al.* eds., Heidelberg, 1991), pp. 385–415.
18. J. P. Ivey, D. M. Davies, V. Morgan and G. P. Ayers, *Tellus*, **38B**, 375–379 (1986).
19. M. R. Legrand and R. J. Delmas, *Atmos. Environ.*, **18**, 1867–1874 (1984).
20. H. Hwang and P. K. Dasgupta, *Anal. Chim. Acta*, **170**, 347–352 (1985).
21. A. L. Lazrus, G. L. Kok, S. N. Gitlin and J. A. Lind, *Anal. Chem.*, **57**, 917–922 (1985).
22. G. Piccardi , R. Udisti and E. Barbolani, *Annali di Chimica (Rome)*, **79**, 701–712 (1989).
23. R. Udisti, E. Barbolani and G. Piccardi, *Annali di Chimica (Rome)*, **81**, 325–341 (1991).
24. R. Udisti, S. Bellandi and G. Piccardi, *Fresenius J. Anal. Chem.*, **349**, 289–293 (1994).
25. G. Piccardi, R. Udisti and F. Casella, *Intern. J. Environ. Anal. Chem.*, **55**, 219–234 (1994).
26. R. Udisti, E. Barbolani and G. Piccardi *in progress.*
27. G. Piccardi, F. Casella and R. Udisti, *Intern. J. Environ. and Chem.*, **63**, (1996).
28. G. Piccardi, R. Udisti, S. Bellandi and E. Barbolani, in: *CNR—PNRA—Proceedings Environmental Impact in Antarctica*—Rome, 1990, pp. 55–62.

29. R. Udisti, F. Casella and G. Piccardi, in: "*Dimethylsulphide: Oceans, Atmosphere and Climate*", (G. Restelli and G. Angeletti Eds., 1993,ECSC, EEC, EAEC, Brussels and Luxembourg, Printed in Netherlands) pp. 153–162.
30. G. Piccardi, E. Barbolani, S. Bellandi, F. Casella and R. Udisti, *Terra Antartica, Antarctic Earth Science Newsletter*, **1**, 134–137 (1994).
31. Y. Gjessing, *Atmos. Environ.*, **18**, 825–830 (1984).
32. Y. Gjessing, *Atmos. Environ.*, **23**, 155–160 (1989).
33. P. A. Mayewsky, M. J. Spencer, W. B. Lyons and M. S. Twickler, *Atmos. Environ.*, **21**, 863–869 (1987).
34. R. Mulvaney, G. F. J. Coulson and H. F. J. Corr, *Tellus*, **45B**, 179–187 (1993).
35. P. Pettre, J. F. Pinglot, M. Pourchet and L. Reynaud, *J. Glaciol.*, **32**, 486–500 (1986).
36. M. Legrand and R. Delmas, *Ann. Glaciol.*, **7**, 20–25 (1985).
37. M. E. Hawley, J. N. Galloway and W. C. Keene, *Water, Air and Soil Poll.*, **42**, 87–102 (1988).
38. H. Sakugawa, I. R. Kaplan, W. Tsai and Y. Cohen, *Environ. Sci. Technol.*, **24**, 1452–1462 (1990).
39. A. Neftel and K. Fuhrer, in: *NATO ASI Series. Vol.* **17**—*The Tropospheric Chemistry of Ozone in the Polar Regions*. (H. Niki and K. H. Beker eds., Heidelberg, 1993) pp. 219–233.
40. K. Fuhrer, A. Neftel, M. Anklin and V. Maggi, *Atmos. Environ.*, **27A**, 1873–1880 (1993).
41. M. O. Andreae and H. Raemdonck, *Science*, **221**, 744–747 (1983).
42. G. E. Shaw, *Rev. Geophis.*, **26**, 89–112 (1988).
43. B. C. Nguyen, N. Mihalopoulos, J. P. Putaud, A. Gaudry, L. Gallet, W. C. Keene and J. N. Galloway, *J. Atmos. Chem.*, **15**, 39–53 (1992).
44. M. Legrand, C. Feniet-Saigne, E. S. Saltzman, C. Germain, N. I. Barkov and V. N. Petrov, *Nature*, **350**, 144–146 (1991).
45. M. Legrand and C. Saigne, *Atmos. Environ.*, **22**, 1011–1017 (1988).
46. M. R. Legrand, R. J. Delmas and R. J. Charlson. *Nature*, **334**, 418–420 (1988).
47. E. S. Saltzman, in: NATO A.R.W.—Ice core studies of global biogeochemical cycles—Annecy (F) 1993.
48. D. L. Savoie and J. M. Prospero, *Nature*, **339**, 685–687 (1989).
49. R. J. Delmas, *Nature*, **299**, 677–678 (1982).
50. R. J. Delmas, M. Legrand, A. J. Aristarain and F. Zanolini, *J. Geophys. Res.*, **90–D7**, 12901–12920 (1985).
51. R. J. Delmas, in: *NATO ASI Series. Vol.* **G6**—*Chemistry of Multiphase Atmospheric Systems* (W. Jaeschke ed., Heidelberg, 1986), pp. 249–266.
52. M. R. Legrand and R. J. Delmas, *Nature*, **327**, 671–676 (1987).
53. M. H. Herron, *J. Geophys. Res.*, **87–C4**, 3052–3060 (1982).
54. D. Wagenbach, U. Gorlach, K. Moser and K. O. Munnich, *Tellus*, **40B**, 426–436 (1988).
55. J. M. Prospero, D. L. Savoie, E. S. Saltzman and R. Larsen, *Nature*, **350**, 221–223 (1991).
56. G. P. Ayers, J. P. Ivey and R. W. Gillet, *Nature*, **349**, 404–406 (1991).
57. T. S. Bates, J. A. Calhoun and P. K. Quinn, *J. Geophys. Res.*, **97–D9**, 9859–9865 (1992).
58. E. S. Saltzman, D. L. Savoie, J. M. Prospero and R. G. Zika, *J. Atmos. Chem.*, **4**, 227–240 (1986).
59. D. L. Savoie, J. M. Prospero, R. J. Larsen and E. S. Saltzman, *J. Atmos. Chem.*, **14**, 181–204 (1992).
60. W. C. Keene, A. A. P. Pszenny, J. N. Galloway and M. E. Hawley, *J. Geophys. Res.*, **91-D6**, 6647–6658 (1986).
61. F. Maupetit and R. J. Delmas, *J. Atmos, Chem.*, **14**, 31–42 (1992).
62. C. Saigne and M. Legrand, *Nature*, **330**, 240–242 (1987).
63. C. Saigne, S. Kirchner and M. Legrand, *Anal. Chim. Acta*, **203**, 11–21 (1987).
64. G. P. Ayers and J. L. Gras, *Nature*, **353**, 834–835 (1991).

NEW ASSESSMENTS ON CFCs TROPOSPHERIC CONCENTRATION LEVELS MEASURED IN TERRA NOVA BAY (ANTARCTICA)

F. BRUNER, M. MAIONE and F. MANGANI

Università degli Studi di Urbino, Istituto di Scienze Chimiche, 6, Piazza Rinascimento, 61029 Urbino (PS), Italy

Data concerning tropospheric levels of some chlorofluorocarbons and other halogenated hydrocarbons are reported. These are referred to atmospheric samples collected during the 1993/94 Austral Summer Antarctic Campaign in Terra Nova Bay. Samples collected in Stainless Steel canisters were analyzed in our laboratory by gas chromatography-electron capture detection (GC-ECD). A comparison with data obtained in previous campaigns is reported as well.

KEY WORDS: Chlorofluorocarbons, atmosphere, Antarctica, GC-ECD.

INTRODUCTION

As already stated in some previous papers[1–4], the intent of this research work is to give data on the tropospheric levels of selected man-made C_1-C_2 halogenated hydrocarbons, whose fundamental role in the Global Change is well known. Compounds that are object of this study are: $CFCl_3$ (F11), CF_2Cl_2 (F12), $C_2F_3Cl_3$ (F113), $C_2F_4Cl_2$ (F114), CH_3CCl_3 and CCl_4. Their involvement in the Global Change is a double one: they are responsible both for stratospheric ozone layer depletion, and for "greenhouse effect". The capability of these compounds to destroy stratospheric ozone has been widely demonstrated[5–7]. As a consequence, they have been listed in the "Montreal Protocol"[8] as the most dangerous substances for the ozone layer. The Protocol, and subsequent amendments, required developed countries to eliminate CFCs production by 1996.

The involvement of CFCs in the "greenhouse effect" is due to their strong ability to absorb, in the range of the "atmospheric window", IR radiation emitted by the earth surface, even though their atmospheric levels are relatively low if compared with those of major "greenhouse gases" (CO_2 and CH_4). Furthermore, their percentage annual growth is much greater than that of CO_2 and CH_4. On this basis, it could be hypothesized that, by the end of this century, CFCs could become the main responsible factor, after CO_2, for the "greenhouse effect"[9].

For these reasons, data concerning tropospheric levels of the aforementioned compounds in a remote region, distant from their major emission sources, are very important. In fact, the average global concentration of these compounds can be calculated, and their secular trend of growth rates estimated. In this frame, Antarctica can be positively considered as the ideal site for these kind of measurements.

In the present paper, data concerning the 1993–94 Austral Summer Antarctic Campaign, which took place within the Italian Research Programme in Antarctica, are reported, and they are compared with those obtained during the past campaigns. Furthermore, some modifications have been introduced to simplify the sampling and analytical procedures.

EXPERIMENTAL

Sampling. As the only source of local contamination can be ascribed to the Italian Station activity, the choice of sampling sites was made based on their distance from Terra Nova Bay Station (Ross Sea Region, Northern Victoria Land, Antarctica); all sites were at least 30 Km away from the Italian Station, as shown in Figure 1.

During the 1993–94 Austral Summer campaign air samples were collected in 0.85 and 16 L Stainless Steel canisters[10] (Biospheric Research Corporation, Hillsboro, OR, USA), whose internal walls were electropolished to provide a chromo-nickel oxide skin, which makes inactive the surface. As CFCs are particularly inert, they are stable in canisters for several months.

Air was drawn inside the canisters by means of a portable ultra-clean air pump (model FC-1121, Biospheric Research Corporation, Hillsboro, OR, USA), whose air contact surfaces were all in Stainless Steel, in order to eliminate any release of organic compounds. Gas-tight connections between the canister and the pump were assured by two Nupro valves (Nupro Co., Willoughby, OH, USA) and a purge-tee fitted with a pressure gauge.

In Figure 2 a diagram of the sampling apparatus is shown. Air was drawn inside the canister at a pressure of 3 atm., and flushed several times, to assure a complete washing, before the final sampling. Then canisters were tightly closed and stored at room temperature before the analysis performed in our laboratory.

Sample introduction into the GC unit. Samples, collected with the above described procedure, were analyzed in gas chromatography with an Electron Capture Detection (ECD). As ECD is a very sensitive detector towards halogenated compounds, no preconcentration step is needed when F12, F11, F113, CH_3CCl_3 and CCl_4, whose atmospheric concentrations are adequate to detector sensitivity, have to be determined. Air samples were introduced into the gas chromatographic unit by means of a conventional six-port valve (Valco Instrument Inc., Houston, TX, USA) system, as described elsewhere[11].

When F114, whose atmospheric levels are relatively low, has to be analyzed, a preconcentration step is needed. A measured volume (0.5 L) of air from the canister is passed through a trap made with a glass tubing (10 cm long, 0.3 cm I.D.) filled with 230 mg of Carbotrap 1 60–80 mesh (Alltech, Deerfield, IL, USA). During the enrichment step, due to the low break through volume of F114 at room temperature, the trap is kept at the temperature of –70°C by means of dry ice. After this step, trap is connected to the GC by a four-port valve. When the trap is not in line with the gas chromatograph, it is heated in a home-made oven for two minutes at 270°C, then the valve is switched and the desorbed analytes are transferred by the carrier gas into the GC column[12–13].

Gas chromatographic analysis. The GC-ECD unit was a DANI 8521-a (DANI Strumentazione Analitica SpA, Monza, Italy) equipped with a glass column (2 m long,

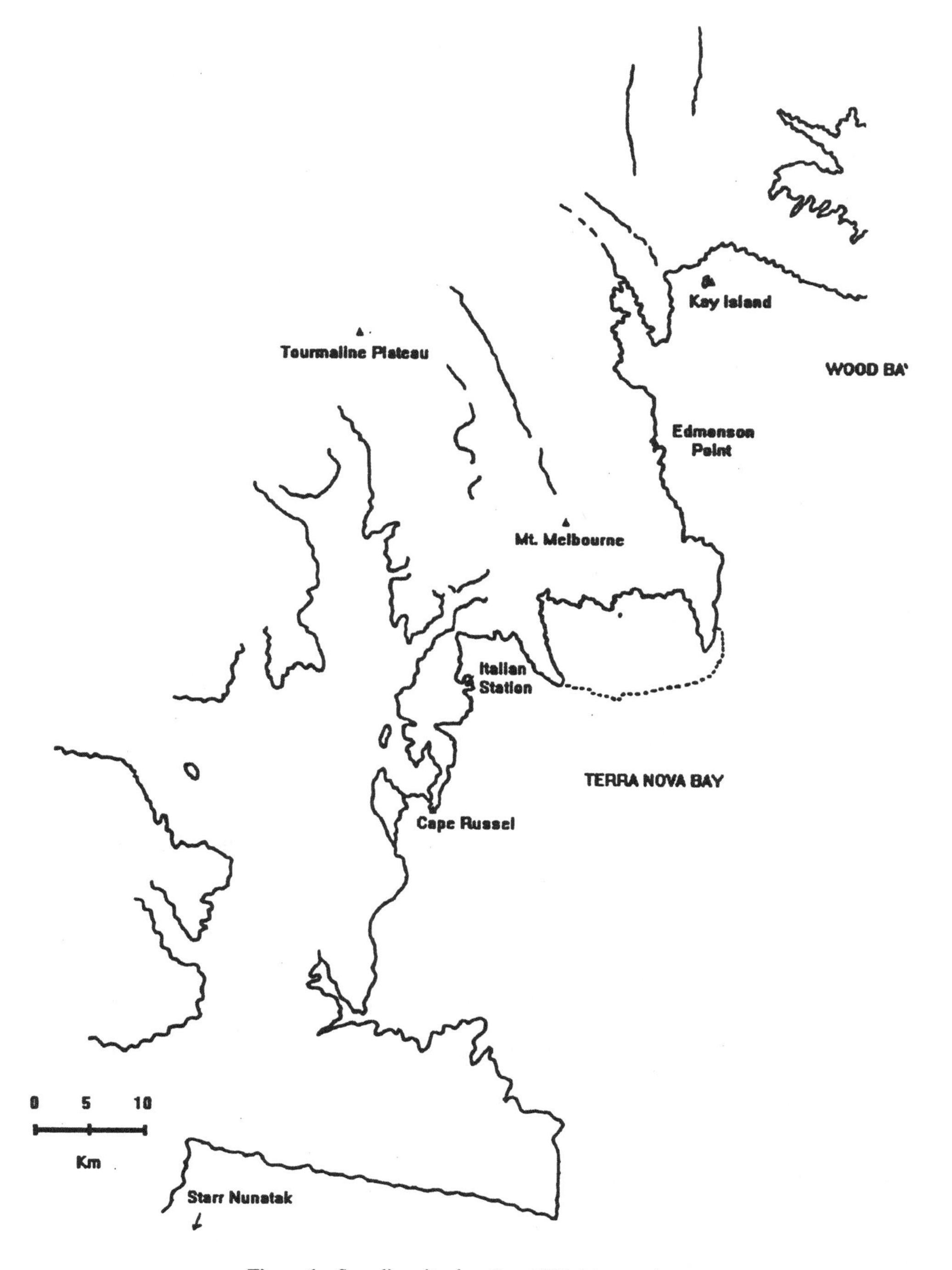

Figure 1 Sampling sites location, 1993–94 campaign.

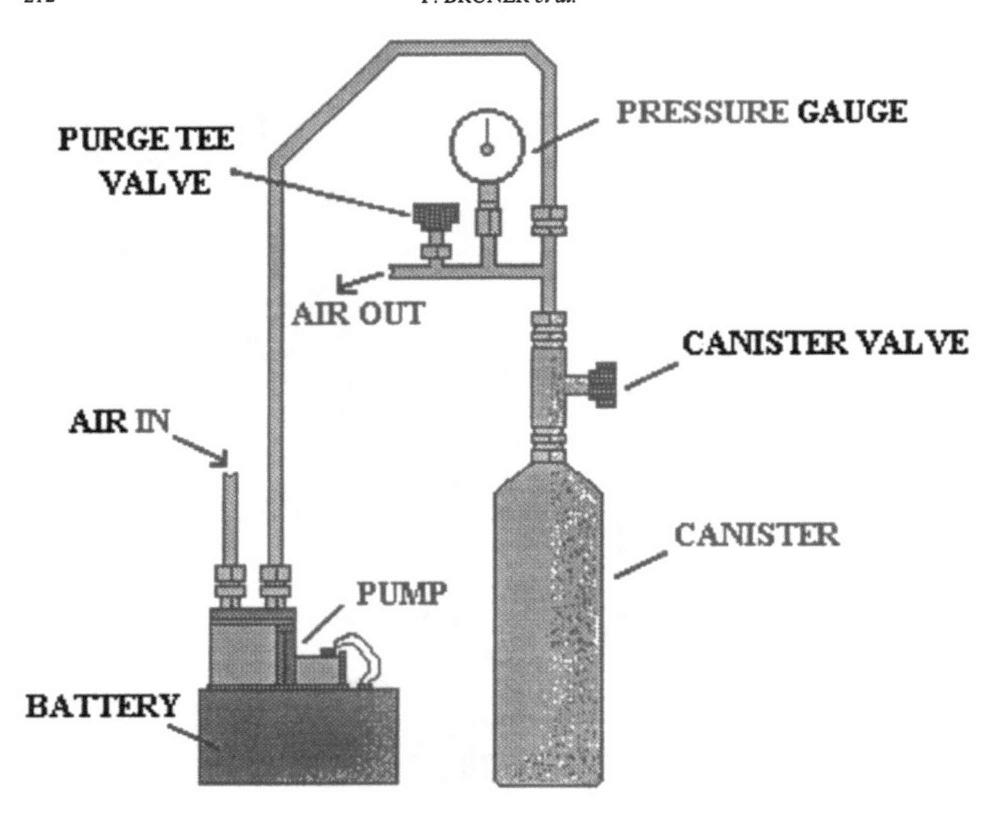

Figure 2 Diagram of the sampling apparatus.

0.2 cm I.D.) packed with Carbograph 1 (Alltech, Deerfield, IL, USA) coated with 1% SP1000 (Supelco, Bellefonte, PA, USA). The temperature programme was: 4 min. at 40°C, then 4°C/min to 120°, hold 1 minute, then 20°C/min to 180°C, hold 15 minutes. UHP Nitrogen was used as carrier gas.

In Figure 3 a chromatogram obtained analyzing a 2.5 mL air sample collected in Antarctica with the direct injection procedure is shown.

Quantification was performed by comparing the actual sample chromatogram with one obtained analyzing a 2.5 mL of a standard mixture of the compounds of interest generated by means of home-made Teflon FEP permeation tubes[14]. The concentration of each compound in the standard solution is close to that of the actual sample, and anyway within the linear range of the calibration curves.

RESULTS AND DISCUSSION

In the past years, our group had set up a sophisticated analytical procedure[12,13] for the analysis of CFCs in the atmosphere of both industrialized and remote areas. This procedure makes use of a high resolution mass spectrometer used in the selected ion monitoring (SIM) mode as a specific detector, and of thermally desorbed traps for

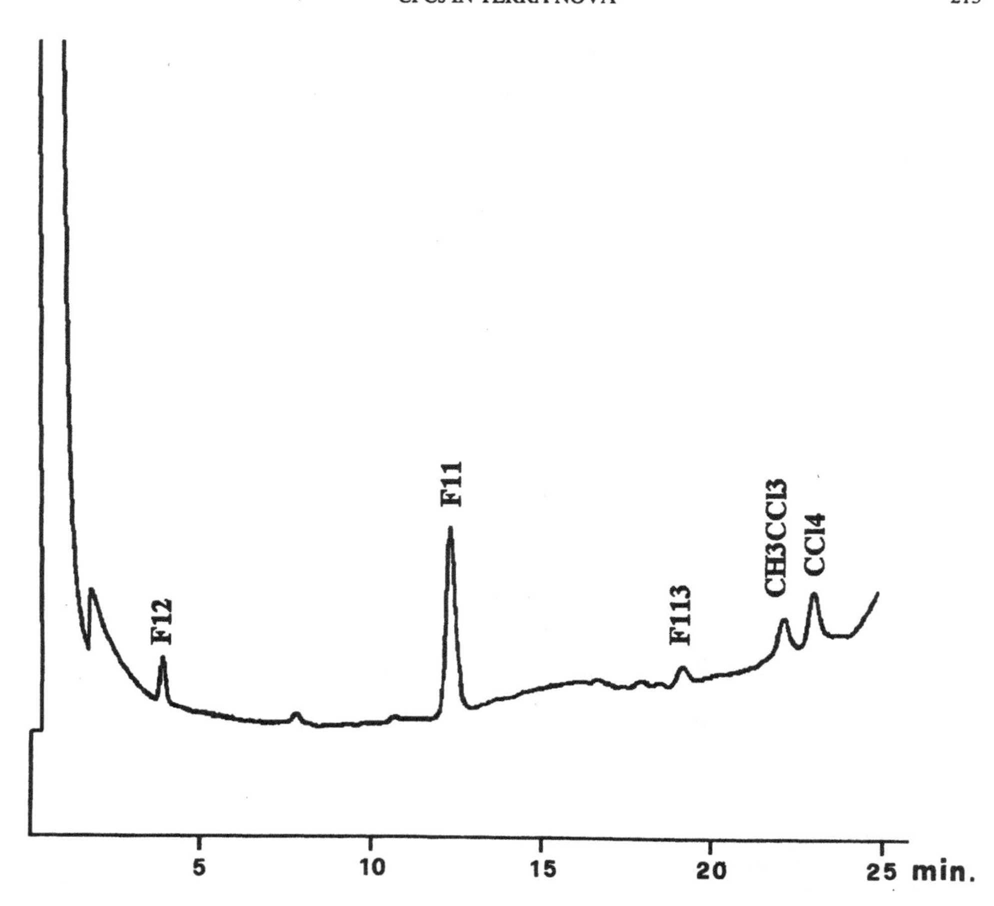

Figure 3 Gas chromatogram obtained after direct injection of 2.5 mL of air collected in Antarctica during the 1993/94 campaign. For chromatographic conditions, see text.

sampling. In this way trace compounds[15] could be detected and interferences due to oxygenated or double bonds containing compounds could be eliminated.

After several years of experience in remote areas[1–4], we decided, due to the purity of Antarctic atmosphere, and on the basis of comparative results obtained using MS and EC detection, that the analysis in remote areas, such as Antarctica, could be performed by using EC detection and canister sampling.

The canister sampling procedure is much easier than the one which makes use of enrichment traps, and this fact is not negligible when operating in severe climatic conditions. Furthermore, each sample could be analyzed several times in order to test the repeatability of the overall analytic method. In this case at least six samples (2.5 mL volume) from each canister were analyzed. Finally canisters showed to be positively inert towards most of the compounds of interest for a relatively long time, meanwhile contamination problems might occur when storing enrichment traps for several weeks.

In Table 1, data obtained analyzing samples collected in six different sampling stations during the 1993–94 Antarctic Campaign are reported. As expected, no appreciable differences in CFCs concentration levels in the different stations were found.

Table 1 Halocarbon concentrations measured in six different sampling sites expressed in ppt (10^{-12} v/v).

Station	*F12*	*F114*	*F11*	*F113*	*CH_3CCl_3*	*CCl_4*
Tourmaline Plateau 74°08' S–163°26'E	524 ± 7%	26 ± 10%	312 ± 9%	97 ± 9%	149 ±14%	108 ±15%
Mount Melbourne 74°20' S–163°20' E	522 ± 6%	24 ± 9%	313 ± 8%	86 ± 8%	182 ± 12%	114 ± 16%
Cape Russel 74°55' S–163°50' E	516 ± 6%	23 ± 11%	292 ± 4%	80 ± 9%	152 ± 11%	130 ± 12%
Starr Nunatak 75°54' S–162°33' E	504 ± 5%	27 ± 12%	313 ± 5%	98 ± 11%	188 ± 10%	102 ± 17%
Edmonson Point 74°20' S–165°07' E	502 ± 9%	21 ± 10%	305 ± 10%	81 ± 12%	150 ± 16%	105 ± 14%
Kay Island 74°04' S–165°19' E	503 ± 3%	22 ± 9%	311 ± 7%	80 ± 7%	161 ± 13%	112 ± 18%

In Table 2, mean values are reported together relative standard deviations calculated on six samples.

In Figure 4 the plotting of F12, F11, F113 and CH_3CCl_3 concentrations measured during four past Antarctic campaigns against time (years) is shown. The gap in the record during 1991–93 was due to a lack of air samples.

These plots show that, at the moment, a flattening of the concentration levels ascribed to the restriction policies concerning CFCs production and consumption adopted by developed countries further to the "Montreal Protocol", is not so evident as it is in the northern hemisphere.

Data concerning concentration levels of compounds like F14 and F113, whose measurements in the southern hemisphere are rare, are particularly interesting. Anomalous data concerning carbon tetrachloride levels could probably be ascribed to the fact that this compound is the most reactive, and some decomposition problems might occur.

Table 2 Mean halocarbon concentrations in Antarctic trosposphere, 1993/94 campaign.

Compounds	*ppt (10^{-12} v/v)*
F12	512 ± 2%
F114	24 ± 1%
F11	307 ± 2%
F113	87 ± 2%
CH_3CCl_3	163 ± 2%
CCl_4	111 ± 2%

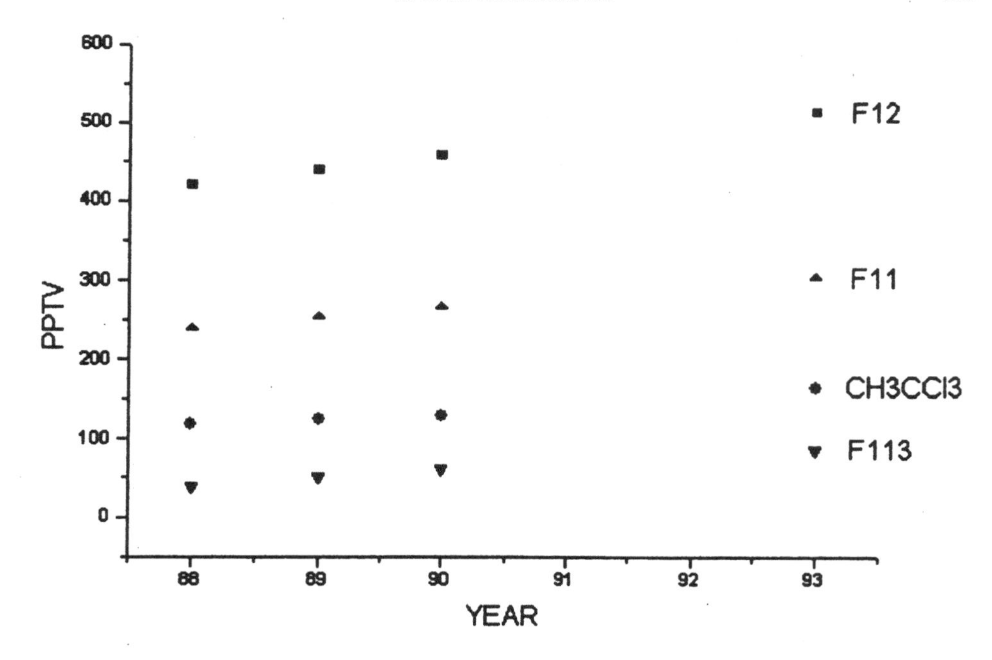

Figure 4 Selected halocarbons concentration levels plotted against time (years).

Acknowledgements

This work was financially supported by the Italian National Programme for Antarctic Research (PNRA).

References

1. G. Crescentini, M. Maione and F. Bruner, *Ann. Chim. (Rome)*, **81**, 491–501 (1991).
2. G. Crescentini, M. Maione, F. Mangani, E. Sisti and F. Bruner, *Italian Research on Antarctic Atmosphere* (3rd Workshop Porano, October 1990, Conference Proceedings, M. Colacino, G. Giovannelli and L. Stefanutti, Eds., Bologna, Italy 1992) pp 91–98.
3. G. Crescentini, M. Maione and F. Bruner, *Italian Research on Antarctic Atmosphere*, (4th Workshop Porano, October 1991, Conference Proceedings, M. Colacino, G. Giovannelli and L. Stefanutti, Eds., Bologna, Italy 1992) pp 205–213.
4. F. Bruner, G. Crescentini, M. Maione and F. Mangani, *Intern. J. Environ. Anal. Chem.*, **55**, 311–318 (1994).
5. M. J. Molina and F. S. Rowland, *Nature*, **249**, 810–812 (1974).
6. F. S. Rowland and M. J. Molina, *Rev. Geophys, Space Phys.*, **13**, 1–35 (1975).
7. M. J. Molina, T. L. Tso, L. T. Molina and Y. F. C. Wang, *Science*, **238**, 1253–1257 (1987).
8. *Montreal Protocol to Reduce Substances that Deplete the Ozone Layer, Final Report* (U.N. Environmental Programme, New York, 1987).
9. G. Brasseur in *Climatic Change and Impacts: A General Introduction*. (CEC-Directorate General, Science, Research and Development Report No. EUR 11943, R. Fantechi, G. Maracchi and M. E. Almeida-Texeira, Eds., Brussels 1991).
10. R. A. Rasmussen and J. E. Lovelock, *J. Geophys Res.*, **88 (C13)**, 8369–8378 (1983).
11. F. Mangani, A. Cappiello, B. Capaccioni and M. Martini, *Chromatographia*, **32**, 441–444 (1991).

12. F. Bruner, G. Crescentini, F. Mangani, E. Brancaleoni, A. Cappiello and P. Ciccioli, *Anal. Chem.*, **53**, 798–801 (1981).
13. G. Crescentini, F. Mangani, A. R. Mastrogiacomo, A. Cappiello and F. Bruner, *J. Chromatogr.*, **280**, 146–151 (1983).
14. G. Crescentini, F. Mangani, A. R. Mastrogiacomo and F. Bruner, *J. Chromatogr.*, **204**, 445–451 (1981).
15. G. Crescentini and F. Bruner, *Nature*, **279**, 311–312 (1979).

INDEX